SMART GRID SECURITY

An End-to-End View of Security
in the New Electrical Grid

SMART GRID
SECURITY

An End-to-End View of Security in the New Electrical Grid

Gilbert N. Sorebo and Michael C. Echols

CRC Press
Taylor & Francis Group
Boca Raton   London   New York

CRC Press is an imprint of the
Taylor & Francis Group, an **informa** business

CRC Press
Taylor & Francis Group
6000 Broken Sound Parkway NW, Suite 300
Boca Raton, FL 33487-2742

© 2012 by Taylor & Francis Group, LLC
CRC Press is an imprint of Taylor & Francis Group, an Informa business

No claim to original U.S. Government works

Printed in the United States of America on acid-free paper
Version Date: 20111028

International Standard Book Number: 978-1-4398-5587-4 (Hardback)

Visit the Taylor & Francis Web site at
http://www.taylorandfrancis.com

and the CRC Press Web site at
http://www.crcpress.com

Dedication

For our parents...

Contents

Foreword

The development of a twenty-first century electric grid that has the potential to transform our economy, secure our critical infrastructure from cyber intrusion, and re-fuel our transportation sector is upon us. To realize this vision, we must begin to change how we look at the system—from one made up of component parts to a single grid, where interdependencies between resources are recognized, managed, and efficiently exploited to ensure reliable, low-cost, and increasingly clean energy is available to consumers across North America.

Over the past 60 years, we have treated the "grid" as two separate systems: the demand side, including distribution infrastructure, and the supply side, including transmission infrastructure. Providing the data and visibility between these two systems, including managing the two-way flow of energy and information across multiple interfaces, is the "Smart Grid." These systems stretch from synchro-phasors on the transmission system to smart appliances in the home. This critical link completes the picture of a fully integrated system without boundaries—from end to end; that is, generation to consumption, including all the complex paths electrons might take to get from one edge to the other.

Unlike the distinct systems of the past, where issues affected the demand and supply sides separately, the end-to-end system of the future faces common challenges and drivers. One of the most important of these is *cyber security*. The potential for an attacker to access the system extends from meter to generator, and the responsibility for securing the grid is held equally by asset owners and technology developers across the grid. The security of SCADA systems at the bulk system level is equally as important as distribution-level metering systems that may, for example, control remote service disconnect for millions of customers. Cyber security is a critical issue that must be part and parcel of our Smart Grid and demand response development strategy.

Our society has come to a fork in the road. We have three options available to us. We can abandon the journey and choose to turn back or reconsider the journey to a new electric system, or we can choose to move forward down one of the roads before us. One road appears to be a short-cut. It simply veers off slightly to the right and is most in line with the road that we are currently on. It is the easier choice and promises the shortest route. This is the road of accelerated technology deployment, the dash toward what we think is the Smart Grid. The other road would require some maneuvering to join and is potentially a longer road. This is the road of learning, one that embraces technology deployments in a measured fashion focused on learning and addressing challenges early in the process. This book sets the stage for considering how cyber security should factor into our decision as to what road is best suited for journey.

Cyber security is another significant risk to the system. One of the most concerning aspects of this challenge is the cross-cutting and horizontal nature of networked technology that provides the means for an intelligent cyber-attacker to impact multiple assets at once, and from a distance. The majority of reliability risks that challenge the bulk power system today result in probabilistic failures that can be studied and accounted for in planning and operating assumptions. For cyber security, we must recognize the potential for simultaneous loss of assets and common modal failure in scale in identifying what must be protected. This is why protection planning requires additional, new thinking on top of sound operating and planning analysis.

There are many activities that are performed every day where the knowledge of the risk entailed in doing the activity is completely unknown. The production, transmission, and use of electricity do not constitute one of these activities. There is considerable understanding of the risks associated with the production, transmission, and use of electricity. When devices fail, weather moves through, and unforeseen events take place, the electric grid operators respond by rebalancing the grid to compensate for the event. This is permitted by the design of the grid.

These challenges are the physical challenges to the electric grid. There is significant knowledge of the mean time between failures for mechanical devices. Knowledge of the patterns of outages caused by weather can almost be predicted. The occurrences of the substation vandal, the unforeseen trip of a generator, or many other actions can be managed due to the way the system is either designed or operated.

With planning criteria that ensure the system can handle a credible contingency and operating requirements to reconfigure to handle another, the grid has the necessary robustness to deal with probabilistic risks. In the physical realm, this construct has merit and has been validated because there is a huge statistical basis of knowledge of equipment failure, equipment malicious operation, acts of nature, and other physical world events. There is substantial knowledge of how to practice in and deal with these types of risks to reliability.

With the new era of ever-increasing digital reliance and system complexity, we see the emergence of common vulnerabilities within the computational backbone of the power system that can result in credible contingencies and maximum contingencies, due to common modal failures or coordinated cyber-attacks. This may significantly challenge the ability to rebalance the system.

This fundamental difference between probabilistic risk and risk introduced by an intelligence adversary (or adaptive threats) leads to the conclusion that there really is not enough understanding of all the cyber security issues and impacts that are possible on the electric grid. Indeed, there really is no statistical norm for the behavior of cyber-attackers and information systems and components failure, and their potential impacts on grid reliability.

Finally, in the computational realm that underlays the cyber-framework of the emerging modern grid, there are very targeted and highly structured threats that can effectively impact many systems at once. As we see in business and home computer systems we use daily, the common components of the computers (such as the operating systems, or the hardware, or even applications) can be exploited. As this computer technology moves further into the operational and control components of the electric grid, it is likely that the impacts from an exploit of a common item, be it hardware or application, can quickly outstrip our planning criteria designed for actions in the physical realm.

This book covers the important technical challenges and forces that will shape how we achieve a secure twenty-first-century electric grid. Chapter 2 devoted to standards is an important one as it correctly identifies how they will shape utility behavior and influence the direction of the technology that will be needed over the next 40 years.

I believe that properly developed technical standards will play an important role in establishing a strong foundation for future electric system reliability and security. I also recognize the growing desire, as significant investments are already being made, to adopt standards that will shape Smart Grid technologies. We must achieve these important goals, but I caution against allowing haste to overcome a deliberate and extensive review of these important guides.

A successful standard must demonstrate that, if implemented in a prudent manner, it will result in outcomes that will not adversely affect the reliability or cyber security of the system, whether in part or in whole. Thinking through the real-world outcomes of proposed standards requires that many minds come to the table—from those who design the technology, to those who implement it, to those who must secure it.

The existing standards identify worthwhile technology targets that will certainly enhance efficiency and enable greater flexibility. These benefits, however, also introduce security concerns as critical functions and components would share a common network, common naming, and automatic point configuration; would rely on peer-to-peer messaging;

and would thus be more susceptible to data storms, setting changes, and malicious programming.

This book begins the call to inspire technologists of today and tomorrow to consider the Smart Grid as a tremendous opportunity to create and solve challenges that affect our entire society. Greater involvement by various domain security experts is needed to highlight areas of concern with the first generation of deployed technology, as well as offer potential solutions.

We should continue to seek progress, but also recognize the need to close the gaps in the software and system engineering foundations necessary to ensure that new Smart Grid functionality will be secure, safe, survivable, reliable, and resilient.

Michael Assante
President and CEO
National Board of Information Security Examiners (NBISE)
(former chief security officer for NERC)

Preface

The field of cyber security, or information security, or information assurance, depending on your preference, is an interdisciplinary one. It is part computer science, part statistics, part psychology, part law, and a part of many other fields. Moreover, any cyber security practitioner needs to know what he or she is protecting and where to prioritize. Just as someone should not spend $100 to protect something worth $10, one cannot be blind to the other disciplines. That, in part, was the inspiration for this book. We felt that the Smart Grid cannot be secured if the people who apply controls do not understand what they are trying to secure. Those who deliver electricity have the right to expect the people who serve them to know their business, particularly those charged with protecting it. Several folks in the industrial control systems community have decried this lack of understanding. For those who argue that one cannot secure a system without knowing how it works or the consequences of implementing the wrong security, this book is for you. Our goal is to make the Smart Grid and all its "warts" accessible to not only cyber security practitioners, but also to the media, policymakers, regulators, engineers, utility executives, and even consumers to understand the interplay between the automation of the electric grid and security. It is through such understanding that we can dispel the myths and pursue a level-headed strategy that fully considers both the risk of not acting as well as the costs of acting.

Because we appreciate that some Smart Grid domains are more relevant for some than for others, we broke out the chapters to cover distinct parts of the Smart Grid, such as metering, distribution, and transmission that are frequently separated. While it is true that the Smart Grid is often about interconnecting those islands of control, we also recognize that it is important to acknowledge the distinctions and then discuss how they interact. Nonetheless, we begin our journey in Chapter 1 with a broad-based look at the Smart Grid and all its pieces. For those looking for a high-level overview of the Smart Grid and why it is important, this chapter covers that.

In Chapter 2 we take a quick diversion to the legal and regulatory environment, most of which does not currently apply to Smart Grid deployments. The chapter recounts the evolution of security standards and regulations, including the North American Electric Reliability Corporation Critical Infrastructure Protection (NERC CIP) standards and current efforts by the National Institute for Standards and Technology (NIST). The chapter then proceeds to examine just how the Smart Grid may be regulated and what requirements utilities should prepare for. It is a constant reminder of the dominant role that compliance plays with security, particularly in the electricity industry. The remainder of the book will largely look at security from a risk perspective. However, cyber security practitioners and asset owners would be well served by understanding the likely compliance implications of their decisions.

In Chapter 3 we dive into the core of Smart Grid, focusing on possibly the most visible part, the smart meter and associated devices that make up the advanced metering infrastructure (AMI). This is where much of the media attention is focused and where many of the initial challenges surrounding security and privacy will be addressed. The chapter covers many of the technical details about how metering technology is being implemented and the likely threats and vulnerabilities that utilities will face. Similarly, we also emphasize the likely impacts and consider how they fit into the overall Smart Grid architecture. Because everything is interconnected, the implications of a single compromised meter could be much more severe, depending on the ability of the rest of the infrastructure to withstand an attack. We offer practical and cost-effective approaches to securing this architecture without going broke.

Next, in Chapter 4, we address the next most visible part of the Smart Grid: the home area network (HAN). Here we freely acknowledge that there are no silver bullets with respect to security. Because the consumer is largely in control, a utility is constrained by what kinds of security can be implemented. For that reason, we recommend that utilities limit the trust they put in such devices and effectively assume that some will be compromised by either the consumers who often will own the devices or a malicious third party. That means that regular sanity checking of data and limiting reliance on such data for critical functions are imperative.

And then we move on to the traditional elements of the grid that are less visible but where Smart Grid technology may provide utilities with the largest returns on investment (Chapters 5 and 6). Distribution and transmission are the foundation for the delivery of electricity and, consequently, automating those functions in the wrong way could pose the greatest security risks. We discuss the various elements of this sprawling maze of wires, transformers, and capacitors that evolved over more than a hundred years. Some aspects of the Smart Grid are offering some of the first upgrades the distribution and transmission systems have seen in decades. Therefore, our hope is that the security deployed today will provide a foundation to build upon that will resist attacks well into the future.

In Chapter 7 we take a different perspective on the traditional subject of generation as the Smart Grid seeks to revolutionize, and potentially democratize, the generation of electricity. We explore the concept of distributed generation and the complementary area of micro-grids. While the cyber security challenges in this area are still being conceived, it is likely that giving individuals the ability to generate their own electricity and be paid for it is likely to lead to attempts at fraud and manipulation. Like the chapter (Chapter 4) on the home area networks, this chapter discusses ways that utilities can limit the trust they accord to any one generation source and how to build in additional protections to resist attempts to manipulate or disrupt the reliability of grid functions.

Chapter 8 seeks to pull it all together in what is typically called "operations." Traditionally this has been the ongoing maintenance and support of the various aspects of the grid. Under Smart Grid, it is where all the data come together and where data from meters are matched up with information coming from line sensors that is correlated with reports coming from substation automation equipment. The results are systems that better automate the balance of generation and the load demand, improve responses to outages, and more accurately predict future demand. This requires interoperability between equipment from multiple vendors that historically have not integrated well together, and it also means dealing with the cyber security gaps that frequently occur at the junction points between multiple independent systems. For the first time, a customer's interaction with a utility's web portal can result in an action being taken at the meter level with little or no involvement of a human being. For example, if a utility chooses to, a customer's request to discontinue service can be requested over the Internet via a web browser and the power can be switched off automatically via a disconnect switch on the meter with the communication to that effect traversing a customer information system, a meter data management system, a meter head-end system, a neighborhood collector, and eventually the meter. That's a lot of systems that need to work together and be adequately secured for such a service to work successfully. We'll discuss how cyber security can play a leading role in providing that assurance.

In Chapter 9 we begin to look to the future. While changes to distribution and transmission as a result of the Smart Grid may appear incremental, the introduction of energy storage, particularly the use of plug-in electric vehicles (PEVs), is potentially a game changer not only in terms of the environmental consequences but also in terms of our ability to truly make full use of renewable energy. Storage allows us to maximize our use of unreliable energy sources such as wind and solar, and PEVs make such storage economically viable. But like distributed generation, widely dispersed energy storage prevents untold opportunities for fraud and manipulation that cyber security controls must address. We examine such risks and offer possible solutions for a business model that clearly has not fully coalesced yet and may not do so for at least another decade.

Chapter 10 continues our look forward but presents the issues with respect to the consumer, who will be presented with a host of new ways to both receive and generate energy. We will look at where the consumer is likely to interact with the utility in significant and new ways. In many ways, the consumer is really the key catalyst who will determine whether much of what constitutes the Smart Grid succeeds or fails. The consumer's willingness to stay engaged and feel secure will be critical. As has been the case in the past, the consumer is both a potential victim of cyber-attacks as well as the possible aggressor. Consequently, utilities will need to keep up their guard while still seeking to engage their customers and teach them how to protect themselves and the grid that we all use.

And while we can all hope for the best, we know that things will occasionally go wrong. Utilities and their customers will fall victim to cyber-attack. Inadvertent mistakes will lead to outages and possibly worse. Moreover, the automation that the Smart Grid hopes to deliver can turn a minor glitch into a major disaster in the blink of an eye. Therefore, utilities must plan for new scenarios and be skeptical of any solution that completely relies on automation. A smarter grid does not mean that humans are no longer needed. On the contrary, a Smart Grid will require a smarter workforce that understands not only how a downed tree can affect electric reliability, but also how a sophisticated virus could make a Category 5 hurricane seem tame by comparison. Chapter 11 offers useful advice on how to respond to these new types of incidents and leverage new technology to one's advantage.

Finally, a book on Smart Grid security would not be complete without a little speculation on future cyber security challenges (Chapter 12). As is well understood, the threats will continue to evolve as incentives change and the nature of the technology and defenses changes. An earthquake does not look for new modes of destruction once a structure is fortified, but hackers will always look for a new way around the latest defense strategy. In Chapter 12 we discuss new attacks such as differential power analysis and new ways that that grid can be defended, such as better key management and protection. But as we've learned before, security is a process, not a product, and the same principles of sound risk management, organization-wide collaboration, holistic thinking, and a broad base of security controls will ultimately be the best formula for success. It is our hope that this book serves as both a reminder to what we've always known about security and also a useful tool in applying those principles to an area that may be new to some of you and thought provoking to most of you.

Authors

Gilbert (Gib) Sorebo is a chief cyber security technologist and an assistant vice president of SAIC where he assists government and private sector organizations in addressing cyber security risks and complying with legal and regulatory requirements. He has been working in the information technology industry for more than 19 years in both the public and private sectors. In addition to federal and state governments, Mr. Sorebo has done security consulting in the financial services, health care, and electricity sectors. He is currently responsible for coordinating cyber security activities in the energy sector company-wide. He has been the co-lead of SAIC's Smart Grid Security practice where he established the SAIC Smart Grid Security Solutions Center for product security testing and solution development and contributing to a variety of other Smart Grid security research efforts. Additionally, he has led projects involving NERC CIP, Nuclear Energy Institute (NEI) 08-09, and risk-based assessments of electric utilities. He has helped more than a dozen utilities address their Smart Grid security requirements, including the drafting of several Cyber Security Plans required under Department of Energy Smart Grid Grant requirements and oversaw an SAIC team responsible for conducting a cyber security assessment for the nuclear fleet of one of the nation's largest electric utilities based on guidance from the Nuclear Regulatory Commission. He is also a frequent speaker at national security and utility conferences, such as the RSA Security Conference, CSI Annual Conference, Metering America, Autovation, and the FIRST Annual Conference, where he has given talks on information security liability, the Sarbanes–Oxley Act, e-discovery, Smart Grid security, incident response, breach notification, and several other topics. Sorebo holds a law degree from the Catholic University of America, a masters degree in legislative affairs from George Washington University, and a bachelor's degree in political science from the University of Chicago. He can be reached at gilbert.n.sorebo@saic.com.

 **Michael Echols** is a cyber security consultant who specializes in the development and management of utilities' cyber security programs. This includes the development of governance models, policy development, and compliance. Michael has worked in Energy and Utility markets developing and delivering transformational cyber security solutions for critical infrastructure systems. Michael is recognized for his expertise in cyber security compliance and posture analysis for industrial control systems and Smart Grid technologies. He has worked in the public sector as a cyber security officer for the U.S. government, where he has applied security requirements from both NIST and NERC to industrial control systems, in real-world situations. He has led efforts to assess, secure and remediate risk to generation, transmission, distribution, and advanced residential metering systems for the U.S. government and major U.S. energy providers. Michael holds a master's degree in computer information systems from the University of Phoenix and a bachelor of arts in political science from the University of Michigan.

Acknowledgments

As full-time cyber security professionals striving to help customers or an employer build a smarter and more secure grid, finding the time to write this book has not been easy. We did not have the luxury of a sabbatical despite receiving encouragement to deliver this important message. To begin with, we would like to thank our families for the support and encouragement we received during the long nights and weekends devoted to writing.

Individually, there are so many to thank; and no matter how hard we try, we will inevitably miss someone. Nonetheless, it is important that we call out these individuals. First, we want to express our sincere appreciation to Josh Wepman who first saw the need for a focus on Smart Grid security at SAIC. Josh taught us a lot, and it was largely his enthusiasm and commitment to this field that motivated our involvement and ultimately led to writing this book. Similarly, the Smart Grid security team at SAIC deserves our thanks for continuing to challenge our thinking and growing our knowledge. Specifically, Rhonda Blachier, Matt Franz, Cedric Robinson, Bruce Rosenthal, Scott Stables, Daryl Thompson, Lynda McGhie, Tim Walsh, Ben Lindsey, Gerry Gallagher, and Frank Flynn deserve our thanks for their tireless dedication to the Smart Grid security cause at SAIC. We have learned a great deal from all of you. Additionally, we wish to thank Annabelle Lee for her courageous leadership in assembling and directing a government, industry, and academic team of over 400 in advancing thought leadership and guidance to assist utilities, vendors, and integrators in securing the grid. This book draws heavily from those efforts. Additionally, Mike Ahmadi has been a valuable resource and confidant in this learning experience. He truly inspires all of us.

A great deal of knowledge has been instilled in us as a direct result of special relationships with experts in the utility field. We have had the privilege of working with technicians, engineers, and management staff from multiple utility companies. Steve Yexley provided us with great context and understanding associated with the field aspects of a utility. We owe a lot of our cyber security knowledge of control systems to Neil

Mcinnis, who is a true expert in this field. To further this end, a great wealth of knowledge was gained as a direct result of a relationship with Chuck King, who was instrumental in teaching us about how the power system really works. Joe Weiss has provided context in regard to how power generation works. Steve Cobb, who maintains a wealth of knowledge across the entire energy industry, has also contributed through his teachings and helped us understand why things need to happen so that we could figure out what the security impacts are that are associated with our nation's energy-critical infrastructure. Additionally, Randy Dreiling, Mike McElhaney, Darrick Moe, James Potts, Brent Sessions, Subhash Paluru, and Matthew Miller all helped in framing our ideals and relating them to the cyber aspects of the overall power grid. Josh Axelrod has also played a role in the development in this book through his compliance knowledge and power generation expertise. It was through Josh that we were able to appropriately frame our thoughts around how to realistically secure specific aspects of a Smart Grid as well as demonstrate the value of compliance. Robert Still also played a major role in the development of our understanding related to Smart Grid security. Thomas Johnson, Gary Burwasser, Angela Scheibel-Benge, and others have all played a major role in challenging our ideals and encouraging us to do better. And finally, we owe a great deal of gratitude to Laurent Webber, who is Mike's mentor and friend, and who was always there to second-guess our opinions and encourage us to take on the most difficult challenges.

Finally, this book certainly could not have been possible without the help in editing and writing we received. Mike Assante, an acclaimed cyber security general in the battles faced by utilities every day, was gracious enough to write our foreword, and for that we are eternally grateful. Louis Szablya deserves praise for the insightful advice and edits he offered for the distribution chapter. Roz Reece offered valuable assistance in the edits she provided as the gatekeeper of SAIC's brand, a role she performs marvelously despite impossible deadlines and impatient co-workers. And last but not least is the editorial team at Taylor & Francis. Mark Listewnik has put up with our delays and adjustments patiently and supportively. He believed in our vision and helped to see it through for us. And Kat Younce also provided insightful assistance as we struggled to understand the publishing process for the first time. Her advice was always right on the mark and extremely helpful.

What Is the Smart Grid, and Why Should We Care about Security?

1.1 DEFINITIONS: THE TRADITIONAL POWER GRID

Before we can seek to understand what it means when we say Smart Grid, an already loaded term with numerous meanings, we first need to understand the electrical grid itself that we're trying to make smarter. "The U.S. power supply network is the largest, most complex machine ever created and engages the most complex enterprise. It involves some 5,000 corporate entities, 100 million customers, four distinct forms of ownership and multiple levels of regulatory oversight."[1] In essence, it is a mechanism to deliver electricity from generation plants to homes and businesses that use the electricity, leveraging long-distance transmission lines that eventually usher power to local distribution grids that step down electricity to a voltage that is usable. Along the way, sensors, switches, capacitor banks, and reclosers use manual and automated controls to properly route the electricity and protect against harm and outages. In particular, special protection systems or remedial action schemes[2] are in place to ensure that disruptions in one part of the electrical grid do not cascade to affect the rest of the grid.

As we will see, much of the grid is assembled in a somewhat jumbled manner that reflects more than a century of additions, tweaks, and workarounds to provide electricity to nearly every home, no matter how remote. Even using the term "grid" implies a level of organization that does not exist. While not completely autonomous, the three major U.S. grids, located in the eastern and western United States and Texas, are by no means centrally controlled, as each generation source, transmission provider, and local distribution organization has some say in the technology used and the processes to be employed. Amazingly, there are few systems, large or small, that are as reliable despite limited resources, ever-growing demand,

and infrastructures that depend on one another, often without any way to enforce consistency or interoperability between their components.

As a contrast to what we will see with the addition of Smart Grid technology, the traditional electrical grid provides limited mechanisms for control or monitoring. However, such a statement is immediately rebutted by electricity industry veterans, who will correctly note that technologies such as supervisory control and data acquisition (SCADA) and distributed control systems (DCS) have provided visibility into and control of grid functions for decades. Nonetheless, these technologies, until recently, tended to focus on key substations and the generation plant, leaving utilities blind to where exactly outages were occurring along a distribution feeder line, the voltage level at home and business, or whether a transformer needed repair or replacement. Special protection systems, defined above, do, however, provide some indication of where faults are occurring, particularly along major distribution feeder lines. However, cost and geographical considerations often make it a challenge to pinpoint faults and outages, particularly as the power flows closer to the end of the line. At a regional level, independent system operators (ISOs) and regional transmission operators (RTOs) have had access to real-time information on the state of the grid through sensors installed at key transmission substations and along high-voltage power lines spanning the country. However, such data provide only a limited view of what is truly going on with the grid, particularly at a local level, and ignores a critical ingredient: interaction with the consumer.

Despite the fact that the ultimate consumers of the electricity dictate the amount of electricity to be generated and delivered, the current grid regards electricity as a bottomless resource that can be used or not used without consideration as to the cost of that demand or who or what will do the generation. With some limited exceptions, electricity generated on the existing grid must be immediately used by a consumer usually within a few hundred miles of the generation plant. That means that the electricity must be dispatchable and cannot be drawn from unpredictable energy resources in most cases. This "always-on" functionality means that it is sometimes taken for granted. For example, consumers do not select power for their lighting or hair dryer like they would a pay-per-view movie. They instead plug in their appliances or turn them on and receive a bill for whatever they use, with little knowledge of what the costs are for the individual uses.

Because this book is about cyber security, it's worth briefly discussing the cyber security challenges of the traditional grid. It is certainly true that information technology (IT)-oriented automation systems are more limited on the traditional grid, depending on how we define it.

We frequently hear about the "air gap" concept that implied that the ordinary enterprise side of the business that managed typical IT resources such as servers and workstations for functions such as human resources, finance, and procurement was physically separated from the operations technology side that is responsible for the generation, transmission, and distribution of electricity. Moreover, the operations side has relied on specialized control systems that were designed for the real-time nature of electricity. One would also note that typical Transmission Control Protocol/Internet Protocol (TCP/IP) networks were less common and more limited, with communications often transacted through dial-up modems and serial communications technology. However, from a cyber security perspective, the system was more secure in that physical access was often required and less secure in that access methods varied and the application of information security best practices was few and far between. The good news was that automated attacks that relied on known architectures that were consistently applied were harder to launch successfully, but attacks on individual components, such as a dial-up modem in a substation, were often easier because they relied largely on security through obscurity. Similarly, an attack on someone's residential electromechanical meter required physical access because there was no communications path to the meter. Launching attacks on thousands of meters from a remote location just wasn't possible.

1.2 DEFINITIONS: WHAT'S A SMART GRID?

The Smart Grid notion was most likely coined by a team working with the U.S. Department of Energy (DOE). However, the DOE is quick to point out that the real goal is for a smarter grid, as the current grid, while needing some improvement, is by no means dumb. Instead, it seeks to extend what intelligence already exists to more parts of the grid, from direct interaction with consumers to substation automation to more accurate information about generation needs.

So then, what makes the Smart Grid so smart? The answer lies in many functions, both real and imagined. After all, Smart Grid is not just a technology; it's a goal to which we aspire. Consequently, many of the future grid's features are still left to be defined. That said, Title XIII of the Energy Independence and Security Act of 2007 highlights ten characteristics of a Smart Grid:

"(1) Increased use of digital information and controls technology to improve reliability, security, and efficiency of the electric grid.
(2) Dynamic optimization of grid operations and resources, with full cyber security.
(3) Deployment and integration of distributed resources and generation, including renewable resources.

(4) Development and incorporation of demand response, demand-side resources, and energy-efficiency resources.

(5) Deployment of 'smart' technologies (real-time, automated, interactive technologies that optimize the physical operation of appliances and consumer devices) for metering, communications concerning grid operations and status, and distribution automation.

(6) Integration of 'smart' appliances and consumer devices.

(7) Deployment and integration of advanced electricity storage and peak-shaving technologies, including plug-in electric and hybrid electric vehicles, and thermal-storage air conditioning.

(8) Provision to consumers of timely information and control options.

(9) Development of standards for communication and interoperability of appliances and equipment connected to the electric grid, including the infrastructure serving the grid.

(10) Identification and lowering of unreasonable or unnecessary barriers to adoption of Smart Grid technologies, practices, and services."

Many of these features figure prominently in the remainder of this book as they represent the key features of Smart Grid and therefore the most likely targets of cyber security attacks.

As we saw above, much of the current electrical grid is an amalgamation of decades of build-outs, patching, and bolt-ons. The objective was simple. It was to build plants to generate electricity based on coal, diesel, natural gas, nuclear, wind, and solar; construct massive transmission lines to bring the electricity to localities; and distribute the electricity locally. Around this infrastructure grew some monitoring at key locations and, along with aggregate measurements of the demand, that dictated whether more or less generation was needed. Beyond that, however, the electrical grid relied on predictable and modest growth in electricity usage and surprisingly resilient components to keep it operating. Outages in the United States and Canada are rare. While this one-way flow of electricity has worked very well under a vertically integrated electric utility model where one company controls the generation, transmission, and distribution for a given customer, the model starts to fall apart when multiple players are involved. If generation can come from multiple locations and multiple companies at any given time, then more coordination is required. Similarly, if commercial or even residential customers want to generate their own power and return some of it to the grid at the distribution level, then we start to understand where the lack of sophisticated measurement and communications capabilities might pose problems. Finally, if some regions find that there is too much demand and not enough supply and there is no easy way to convey that information to customers to adjust their usage, then we are left with the politically unpalatable and potentially dangerous practice of rolling blackouts. What we will find in the following pages is that Smart Grid was not really meant for the problems of today. It was meant to address the challenges of tomorrow. For the past century, the electrical grid has worked extraordinarily well with limited command-and-control capabilities and little customer interaction. The cost of electricity in most of the United States is cheap and outages are manageable. If it were not for a wide variety of demographic, technological, and socioeconomic changes that are about to sweep our world, the conclusion might be that the grid is working fine and no changes are needed.

As Figure 1.1 depicts, the traditional electrical grid is very hierarchical, with generation at the top, transmission in the middle, and distribution at the bottom operating somewhat autonomously. Under the Smart Grid model, traditional generation still has a large role, but it is augmented by distributed generation in the form of wind, solar, and various other customer-owned generation sources that not only generate electricity for the end customers, but can also sell electricity back to the utility. Additionally, communications networks are added to not only support distributed generation, but also to integrate customer interaction into the equation so utilities can both influence behavior and make better planning decisions based on customer choices. This includes technologies such as advanced metering infrastructure (AMI) that allow smart meters on houses and buildings to relay in near-real-time the usage level of that facility. And through appliance-based communications technologies, grid components can even relay appliance-level usage information and receive commands from the utility to alter the behavior or ultimate operation of the appliance during times of heavy electricity usage. In addition to distributed generation and enhanced communications and measurement capabilities, the Smart Grid also envisions the ability to store electricity for later usage via conventional and evolving battery technologies, compressed air, and pumped storage where water is pumped uphill during off-peak periods and then allowed to run downhill during peak periods to generate electricity. One particularly interesting storage solution is the plug-in hybrid electric vehicle (PHEV) that is capable of both generating and storing electricity through a gasoline engine and battery and can also receive its energy from the electric grid by plugging it in using technology already available. The same technology could then be extended to allow these vehicles to store energy for later use by other devices on the electrical grid when not needed by the vehicle. It is this smoothing out of energy usage to maximize the generating capacity that is one of the key features of the Smart Grid.

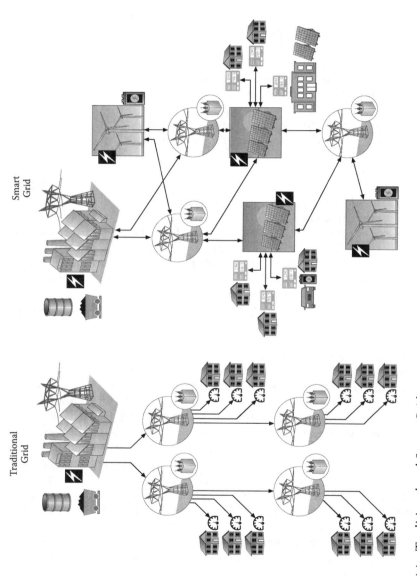

FIGURE 1.1   Traditional and Smart Grids.

1.3 WHY DO WE NEED A SMARTER GRID?

As has been mentioned already, the Smart Grid is not intended to fix problems with the current electrical grid. With a few exceptions in warm climates with heavy population such as Florida and Southern California, electricity is plentiful, reliable, and cheap. However, that is likely to change for a variety of reasons. In his book entitled *Visions for a Sustainable Energy Future*, Mark Gabriel highlights a variety of trends that are affecting our ability to stick with the status quo. They include demographics, the evolution of the energy business, carbon constraints and capacity demands, availability of intelligent infrastructure, and the need for customer engagement. One could argue that the primary driver, however, is economic, and so, if investor-owned utilities cannot make a profit or if electricity prices are too high for consumers, then change will be required. The reality is that multiple factors are driving this economic necessity. Demographic changes have driven more people to warm-weather climates that demand more electricity for air conditioning, usually the largest single electricity cost when it is used for residential purposes. Our electronic culture, from cell phones that need charging to large televisions that draw more electricity than a previous generation, is also increasing demand. And, of course, the supply side is also facing challenges from environmental regulations that restrict traditional fossil fuel-based generation plants, from very long and heavily regulated deployment windows for nuclear, and from the investment costs and long payback period for renewables such as solar and wind. These supply-and-demand challenges alone will likely cause dramatic increases in electricity costs, particularly in large population centers in warm climates. Additionally, electric utilities face an aging workforce with no ready replacements. Both the reality of fewer people to do the work and the preference of the younger generation to rely more on technology are driving utilities to embrace automation and outsourcing, which are not always very well thought out.

Despite being Fortune 500 companies in many cases, large investor-owned utilities face heavy regulation by public utility commissions over what they can charge their customers at the distribution level, and therefore they are constantly searching for greater cost efficiencies and new markets to enter that are less regulated in order to keep investors interested. Taking advantage of the tax benefits, grants, and greater market-orientation available on the generation side are some of the reasons to pursue renewable energy businesses. The DOE Smart Grid grants funded under the American Recovery and Reinvestment Act of 2009 gave some of these utilities an impetus when the economics in favor of Smart Grid investments were still a bit tenuous. More than that, Smart Grid presents a host of business opportunities and technological innovations in an industry that is not historically known for either. Customer engagement, while a bit less compelling, is one of these opportunities, along with the availability of a technology that creates its own demand. What may never have been on the radar of consumers and the public utility commissions that represent them suddenly will become an imperative once easy and useful technologies are available. We never knew we needed cell phones or digital video recorders until they became economical and packaged for consumers. Now they are practically necessities.

The expected benefits for Smart Grid sound a bit anticlimactic when one first looks at the up-front investment. Most experts believe that energy prices will be higher regardless, but through Smart Grid, they will not go up as much. Consumers are likely to experience fewer rolling blackouts as utilities will be able to use demand response programs to turn off selected appliances during peak periods. And consumers will have more control over which appliances are turned off. While Smart Grid promises to reduce some outages through better predictive failure analytics and get help to reduce the duration of outages through near-immediate pinpointing of customers without power, or at least meter connectivity, the Smart Grid will not do much to keep that big old tree from snapping the power lines to one's home during the next storm. Some day, distributed generation and localized energy storage may help mitigate some of the storm-related outages. However, most homes are still likely to have single points of failure where trees and backhoe cuts still pose a danger.

As a business proposition, Smart Grid presents some of the same challenges as cyber security as the savings may be in costs avoided. Given the fairly low electricity rates that most Americans enjoy, the biggest savings will likely come from rates that will not go up as high as they would have otherwise, because as demand for energy increases, some rate increases are inevitable. Consequently, the savings that Smart Grid offers come from cost savings from lower staffing levels through greater automation, smaller revenue losses as a result of shorter outages, and lower generation costs through demand response and dynamic pricing programs that shift usage to lower demand time periods. There are many more Smart Grid use cases that could result in lower costs and fewer revenue losses. However, there are many unknowns. For example, demand response programs depend heavily on customer participation and changes in their behaviors. Savings from greater automation depend on achieving lower employee head counts. And the ability to shorten outages assumes that future outages will be easier to predict and faster to remedy with more advanced but still unproven technology. Added to this is the cost associated with the Smart Grid upgrade. Government subsidies through the American Recovery and Reinvestment Act are helping to ease that burden, but in many cases, utilities are seeking to pass the remaining costs to ratepayers, a proposition that is not going over well with public utilities commissions. Nonetheless, the electrical grid is sorely in need of investment. The only question is whether Smart Grid deployments will allow utilities to delay or avoid replacement of aging infrastructure, or if transformers and other grid workhorses at the end of their lives will demand replacement sooner rather than later.[3]

The future grid is difficult to predict as so many factors come into play. Looking 50 years ahead, one can envision a highly dynamic distribution grid where individual consumers generate their own power through solar and wind and have storage and other resources to draw from when the sun does not shine or the wind is not blowing. Total electric usage will inevitably grow as we grow more reliant on technology and significant loads, such as plug-in hybrid electric vehicles, are added.

1.4 SMART GRID RISKS

While this book is about cyber security risks to the Smart Grid, it is useful to first set the context. As we just saw, the current electrical grid is already facing numerous risks, from physical attacks to failures due to aging infrastructure. The point of this book is not to highlight those risks, but rather to identify the new threats and resulting risks that are driven by the Smart Grid. However, it is worth noting that cyber security attacks are sometimes a precursor to physical attacks, and so, to the extent that Smart Grid technology can be used to increase the probability of a successful attack or to amplify the damage, we will discuss it. Chief among the new threats to the grid is the remote attack and the compromises that could result due to the increasing amount of interconnectivity of data communications networks that control and monitor grid activity. As with any new technology, the promise of increased automation and remote problem resolution also brings us the challenge of protecting an exponentially greater number of attack vectors as each residential meter is a potential entry point in grid communications networks. In Chapter 3, we provide a detailed examination of potential attacks at the meter level.

For now, it is important to understand that rolling out communications functionality to each residence fundamentally changes the threat dynamic and poses some challenges never faced by the electricity industry and, to some extent, never faced anywhere else. This is because a device that the utility has very limited physical control over will be a key component in telling utilities how much electricity to generate and to whom it is delivered. While cable boxes for television and cable modems for Internet access have similar physical control issues, the consequences of widespread compromise are more limited.

Smart Grid cyber security threats are a moving target, and so, determining their likely success is usually more art than science. Throughout this book, we examine the different aspects of the Smart Grid and its vulnerabilities to compromise. There is some evidence that foreign intelligence agencies have already infiltrated the U.S. electrical grid and are passively monitoring it.[4] However, that evidence is speculative at best, based largely on assumptions made about traffic flowing over the Internet targeting a utility. Generally, there is not enough information to identify the nature of the system compromised. For example, a utility's human resources or finance department may become infected with a virus causing no impact on the electrical grid, which is normally physically or virtually segregated in a way that makes attacking it difficult.

This is not to suggest that the current electrical grid or a future version would be less vulnerable. Instead, the underlying premise is that the nature of the attacks, the likely attack vector, and the attacker's motivation are likely to be different from conventional attacks on enterprise networks. For example, attacks on electric meters are unlikely to target the utility's Internet connection. Instead, attackers may either seek to physically compromise a meter first, and then use it to attack the rest of the meter network or target a consumer's Internet connection and use the home's network to reach the meter. As of this writing, the former attack has only been demonstrated in a lab environment using a fair amount of extrapolation, while the latter is only speculative. Moreover, the motivation is not exactly clear. Certainly sabotage is always a concern, but many cyber-attacks require a fair amount of expertise and resources to carry out. In many cases, explosives targeting key substations or transmission lines might be easier to accomplish and more harmful to the grid and electricity consumers. However, cyber-attacks, like those targeting the countries of Georgia and Estonia, can prove more effective when used randomly over time as they create much more uncertainty with respect to the reliability of electric power than a single explosion. Moreover, many attacks can be executed and sustained without ever having to set foot in the United States. This enlarges the potential number of attackers and increases the likelihood that such attacks will be launched when the chances of apprehension are low.

In addition to presenting additional attack vectors, the Smart Grid of the future presents the potential for greater harm if designed incorrectly. As we will see in later chapters, the electrical grid, while incredibly resilient in many ways, was not designed to withstand intentional attacks at its weakest points. Moreover, by creating cascade-like events, the grid's power can be used against it. Where once a coordinated and simultaneous physical attack on numerous substations would be required to cause a massive power outage, a future grid could suffer the same fate through a hacker's keystroke. However, a smarter grid can also be quicker in detecting and responding to such attacks if designed to do so. Like most other technological advancements, the Smart Grid contains the potential for greater efficiency and reliability—but it also has the potential for greater harm.

However, beyond the actual harms and the likelihood that they will occur, there is also the perception challenge. An unfounded belief that the entire grid can be compromised from a single meter can have a powerful impact on a utility's ability to successfully deploy a Smart Grid. Some even speculate that terrorist and other malicious entities often gain as much from behavioral changes resulting from fear and uncertainty as they do from direct attacks. For example, limited attacks on Smart Grid technology can lead people to question the security of broader technology and forestall needed improvements, resulting in higher energy costs and possibly less security. Well-known security expert Bruce Schneier talks about the psychology of security, noting some of the evolutionary influences that cause us to incorrectly estimate risk and to react inappropriately.[5]

When presented with risks associated with new technology that offers benefits that are speculative and far into the future, the natural response is to be skeptical. The notion that a new technology may be vulnerable to hackers in ways that legacy technology is not may cause the public and policymakers to resist the new technology even if legacy technology holds greater risk. Just like electronic voting machines a few years ago, a few well-publicized outages resulting from Smart Grid technology being compromised, or even compelling cases for how it could happen, could be enough to sink the whole effort. Similarly, privacy risks and other harms disproportionately felt by the public, even if small in impact, are likely to result in greater resistance than more significant risks, such as the potential loss of life by a utility employee. Utilities need to work at personalizing the benefits of Smart Grid and effectively explain how risks are mitigated. Otherwise, we could all miss out on some tremendous advances.

So, that raises the question of just how one can sort through all the psychological baggage and accurately identify the real risks. Typically, risk assessment methodologies can be used to identify relevant risks, but this is not without its pitfalls. Quantitative risk assessment methodologies have proven quite useful in the financial services community to evaluate a variety of risks due to the wealth of data available and the high level of transparency inherent in public markets. However, as the recent financial meltdown proved, those models have their limits, particularly when they are built on some faulty assumptions about the broader environment, such as the low likelihood that housing prices would drop significantly over a short period of time nationwide. In information security, this problem is magnified by the fact that most security incidents are not reported, and there is no standardized way of comparing one enterprise to another in terms of risk posture. While car companies can draw on a few metrics such as age, gender, location of residence, and driving record to arrive at a reasonably reliable set of risk factors on which to base premiums, there is no similar consensus that a certain set of controls results in a particular frequency of compromises. Moreover, in all areas where criminal intent is involved, attacks evolve as new controls are added. The most that can be said is that certain industries, such as financial services and national security, tend to be more likely to experience attack. That is reflected in some cyber security insurance policies that exclude large financial institutions from coverage.

For the electricity industry, the question boils down to understanding the core value of its services to its various stakeholders. For shareholders or asset owners, the issue is continued revenue at an acceptable profit. An outage of a single generation plant may not result in a single consumer losing power, but the loss of revenue for the power producer can be significant even if it never makes the news. Conversely, loss of power to a neighborhood of a few hundred people may make the news; and if the problem is not related to weather, it can pose a significant public relations challenge for a utility if the outage lasts for more than a few hours. Ideally, Smart Grid technology should be developed to address those two issues and, through the introduction of advanced sensor technology, to prevent outages and more sophisticated outage management systems to shorten them. However, the risks of centralized control resulting in a centralized malfunction should also be considered. In this brave new world, we may not have the benefit of hindsight to address and minimize the likely risks. Consequently, utilities, vendors, and regulators need to think deeply and objectively to fully understand the kinds of security controls that will best address tomorrow's vulnerabilities and the threats that target them with an eye toward the impacts that are most likely to forestall a move to the Smart Grid or to its demise.

1.5 SMART GRID RISKS VERSUS BENEFITS

With the above risks in mind, it should come as no surprise that a Smart Grid deployment is no guarantee that our electrical grid will be safer, more reliable, or more secure. History has taught us that the adoption of new technology is not always forward-moving. Nuclear power was originally viewed as the wave of the future, but because of incidents like Three Mile Island and Chernobyl, the industry has been stalled for decades even though there is every indication that, when considered in its totality, the technology is likely safer, and certainly cleaner, than fossil fuels. It is only in the past few years that nuclear appears to have regained its footing with the imminent construction of new plants. For Smart Grid, the nuclear comparison is less appropriate as Smart Grid technology is really a collection of various technologies, many of which can work independently of each other. Utilities can choose to automate substations or add new sensors to electrical lines without deploying a single new residential smart meter. As we will see in the following chapters, each Smart Grid element faces its own, often unique, security challenges and may have different stakeholders and stakeholder interests to consider.

However, just as the electrical grid is comprised of areas of both redundancy and interdependency, so too will be the Smart Grid. One of the most important messages the reader can take away from this book is that when deploying an ecosystem like Smart Grid, it is necessary to recognize where to introduce redundancy, resilience, and even self-sufficiency into the existing electrical grid while still preserving the economies of scale that an interdependent grid provides. Doing so incorrectly would mean a grid much less reliable and prone to cascading outages. Doing so correctly means a grid that increases reliability, reduces costs, and makes innovation easier. While this book is about security, the application of a sound, quality system is critical as security is really just a subset of quality. That means giving careful consideration to not only what the bad guys want to do to the grid, but also to what the good guys may do inadvertently. Both quality and security must be integrated into every process and considered right from the beginning. That goes for product vendors, integrators, and end customers. While everyone has his or her role to play, no one can ignore either security or quality.

And that brings us to the question of control and outsourcing. Many utilities frequently ask the question of just what it is they need to buy to be secure or compliant with security requirements. A security professional's typical response is that security is a process and not a product. One could also add that security is not an outsourced service. Ultimately, the stakeholders are responsible for the security of the grid. In some cases, that responsibility is spread throughout the interconnecting pieces of the grid, from distribution-only utilities to transmission providers to generation and even to the consumers. However, within each of those spheres, there is an obligation to provide appropriate security and to ensure compliance with regulations. That does not mean that third parties cannot be brought in to provide monitoring services, to patch systems, and even to take over responsibility for drafting and implementing policies and procedures. However, just as a criminal defendant knows that regardless of how much of the decision making he turns over to his lawyer in putting on his defense, he is ultimately the one who will serve time if there is a conviction. Allowing someone else to take responsibility for security can have serious consequences.

Consequently, any outsourcing or reliance on standards or regulations still requires constant vigilance on the part of the utility. A contractor's main responsibility is to meet the requirements of the contract. Resisting security threats may be part of that responsibility, but rarely are contractors responsible for attacks once they have met their contractual responsibilities. The key to that relationship is typically to define clear metrics for contract compliance that align well with the security risks faced by the organization and ensure that there is flexibility within the contract to adjust metrics and contractor responsibilities with the risk posture changes. So yes, one can outsource security functions, and in many cases probably should, but outsourcing risk is another matter. Even insurance policies, if available in this area, can only pay you cash. They cannot restore customer confidence or correct compliance violations.

Throughout this chapter we have discussed many of the risks to Smart Grid from the 10,000-foot level. We also understand that the benefits of the Smart Grid are far from guaranteed even if security could be assured. As we proceed through an examination of the various Smart Grid technologies and their various security risks, it is important to keep in mind the precarious road on which we travel. Any bump along the way, from a major security event involving Smart Grid technology to an unexpected rate increase, could spell doom for a Smart Grid deployment or at least cause some major delays. What that means is that every piece is important. Utilities cannot afford to mess up security, their Smart Grid advertising campaign, the usability of their in-home devices, the rates they charge, or their quality control mechanisms. They are all important. And so with that in mind, let us look at the security part a little more closely.

ENDNOTES

1. Mark A. Gabriel, *Visions for a Sustainable Energy Future*. Fairmont Press (Lilburn, GA), 2008, p. 62, quoting the Electric Power Research Institute's (EPRI's) "Electricity Sector Framework for the Future."
2. The NERC Glossary of Terms Used in Reliability Standards defines a special protection system (SPS) as "An automatic protection system designed to detect abnormal or predetermined system conditions, and take corrective actions other than and/or in addition to the isolation of faulted components to maintain system reliability. Such action may include changes in demand, generation (MW and Mvar), or system configuration to maintain system stability, acceptable voltage, or power flows. An SPS does not include (a) underfrequency or undervoltage load shedding or (b) fault conditions that must be isolated or (c) out-of-step relaying (not designed as

an integral part of an SPS). Also called Remedial Action Scheme."
Visited online at http://www.nerc.com/files/Glossary_12Feb08.pdf.

3. A transformer has a useful life of 40 years but surveys indicate that the average transformer is 43 years old.

4. Siobhan Gorman, Electricity Grid in U.S. Penetrated by Spies. *Wall Street Journal,* April 8, 2009, visited online at http://online.wsj.com/article/SB123914805204099085.html.

5. Bruce Schneier, The Psychology of Security. Visited online at http://www.schneier.com/essay-155.html.

CHAPTER 2

The Smart Grid Evolution
Smart Grid Standards, Laws, and Industry Guidance

2.1 INTRODUCTION

Smart Grid represents the largest upgrade to the utilities' infrastructure in the history of their existence. Utility companies have long managed and delivered energy through various process control systems. These systems have largely been comprised of assets that relied on serial communications for interconnectivity and copper-based communications for remote access. In most cases, however, these systems were operated manually by trained technicians, independent of the technology. As Smart Grid technology is rolled out, however, these systems are relying more and more on automation, managed by trained IT (information technology) staff. As a result, we are starting to see convergence between operations IT and corporate IT in order to realize the cost savings of using standardized IT processes to manage sophisticated control system applications. To define Smart Grid more specifically in terms of the operational function it will provide is simply to upgrade legacy architecture to a more sophisticated IT architecture. The problem has been that all the security controls that have been defined and developed for advanced IT systems and services cannot be applied to Smart Grid technology because they will have a significant impact on performance. Moreover, at the core of Smart Grid is the need to maintain reliable systems, in order to ensure that power management remains constant. While upgrading legacy grid technology to advanced Smart Grid technology enhances performance and augments reliability, the introduction of security controls meant to protect this advanced technology many times cannot be integrated into the architecture because they stiffen performance and reliability.

IT security controls, standards, regulations, etc. many times cannot be applied to Smart Grid technology, which illustrates the risks associated with Smart Grid. There are a number of standards that have been developed to augment security for various sectors, such as banks, schools, hospitals, etc. In those cases, traditional access control, identity management, system integrity, communications protection, and other controls are integrated into the architecture in order to prevent external and internal threats from compromising the network and supported systems. In the Smart Grid space, which represents the integration of legacy technology with advanced IT technology, reliability is affected many times when these controls are implemented. The result has been that utilities implementing Smart Grid technology have found it difficult to secure their architectures to the satisfaction of what is generally considered acceptable in the IT world. And without a standardized approach to specifically secure the Smart Grid, risk management has become requiring what can be implemented and forgetting about everything else, meaning that organizations may decide to ignore the controls that cannot be implemented because of reliability concerns and focus instead on what can actually be implemented.

The problem with this approach is that risk still exists and in many cases goes undocumented and thus not analyzed. Like anyone else, utilities seek solutions that can actually be used. A long-standing example is where a human-machine interface (HMI) in a substation is implemented to provide digital control over relays used to manage generation. The legacy method to control the relays is to simply manually turn a switch that opens or closes the breaker to which the relay connects. The problem is that with so many relays to control and so many substations owned, it is not scalable to man every substation and device that is controlled by the utility. This is mostly because of all the cost pressures associated with maintaining an army of technicians qualified to operate such equipment. As IT technology has improved over the years, however, utility vendors have realized that it would be more scalable to implement an HMI to allow technicians to control multiple relays from one computer console in a substation. As the workforce ages and qualified resources become more limited, utilities will find it difficult to find the skilled labor that is needed to operate these sophisticated systems. As a measure to manage this problem, they might decide to fully network the HMIs and begin controlling IEDs through national control centers, where they are able to better pool resources. This transition from manual operation to remote operation over an IT-enabling network represents what Smart Grid is all about. This illustrates that all the vulnerabilities and risks associated with the implementation of these IP networks for HMIs should become concerns for the industry. Smart Grid is then effectively bringing IT risk to the grid. This is primarily because history has shown us that as new technology becomes available, threats will learn them and expose weaknesses in those technologies. Furthermore, the more well known technologies become, the more threats that will concentrate on them.

Smart Grid as a new technology using traditional IT principles represents a new vector of attack for threats. Customization of IT platforms is being done in order to leverage the advanced IT technology in scaling legacy grid processes. This means that traditional operating systems may be used to support the HMI or simply that the HMI is an application developed by a vendor and installed on an operating system. These applications leverage operating platforms for their ability to connect to standardized IP communications networks and support the use of various protocols for communications, meaning that the legacy protocols are simply being modified to run on IP communications networks from application to application. The application running on an operating system that is being called HMI is used to communicate with a relay that understands various legacy protocols. As the operating system does not inherently use these protocols, the application converts what it wants to say to the relay, using the TCP/IP (Transmission Control Protocol) stack, and then is reconverted to the legacy protocol when it reaches the relay, usually from a switch that the relay has a serial connection to. It is this process that enables remote management of legacy relays, for example. There are advanced relays that are being implemented more and more that operate under the same concept as the HMI, essentially an application installed on a traditional operating system, thus mitigating the need for the serial-to-IP conversion at a switch in the middle.

These changes all represent an example of what Smart Grid is and how it is transforming legacy systems into advanced IT systems. From a security standpoint, however, these systems, while leveraging the benefits of IT, are ignoring the security concepts that have been evolving since IT became a popular target for threat sources. For example, the HMIs may not be able to support the use of traditional security controls such as Secure LDAP (Secure Lightweight Directory Access Protocol), antivirus, normal patching cycles from the operating system vendor, etc. This means that they are being deployed on data networks without security controls that would typically be seen as protecting the IT-enabling asset. The rationale associated with this is because the vendor developed the solution to leverage IT and because the legacy portion of the system needs all the available resources of the platform in order to function correctly. To make a long story short, utilities are running into situations where if they implement these security controls, the reliability of their HMIs (in this case) may be compromised. In such cases, this means that cyber security represents an actual cyber-risk to the very Smart Grid technology being developed and deployed. This is the challenge associated with securing Smart Grid technology, and the traditional approach to IT security simply will not work in environments like this for the reasons stated above. As security standards represent the frameworks needed to secure IT assets, and those frameworks represent security controls, how do you secure technology that cannot be secured with traditional security controls? How are security standards changing to reflect this new found reality? And finally, where are we headed, and how will the overall Smart Grid be secured?

2.2 REGULATIONS, SMART GRID, AND THE BULK ELECTRIC SYSTEM

The Federal Energy Regulatory Commission (FERC) is an independent agency that regulates the interstate transmission of electricity, natural gas, and oil.[1] The Energy Policy Act of 2005 provided FERC with additional responsibilities, to include the authority to regulate the transmission and wholesale sales of electricity in interstate commerce, licensing and inspection of private, municipal, and state hydroelectric projects, protecting the reliability of the high-voltage interstate transmission system through mandatory reliability standards, monitoring and investigating energy markets, and enforcing FERC regulatory requirements through imposition of civil penalties and other means. FERC does not specifically regulate retail electricity and natural gas sales to consumers, regulate nuclear power plants, or maintain oversight for problems related to the failures of local distribution facilities. These facts are key to understanding why the North American Electric Reliability Corporation (NERC) only has regulatory authority over the bulk electric system. NERC's primary mission is to ensure the reliability of the North American bulk power system. NERC is the electric reliability organization (ERO) certified by FERC to establish and enforce reliability standards for the bulk power system.[2] "NERC develops and enforces reliability standards; … monitors the bulk power system; and educates, trains and certifies industry personnel."[3]

It is important to understand that NERC is a non-profit corporation established to carry out and execute the regulatory obligations of FERC. This responsibility is vested in the fact that NERC has been established as the ERO, which provides them with the ability to enforce reliability standards. The reliability standards must be approved by FERC and, upon such approval, NERC is able to execute its mission. The NERC Critical Infrastructure Protection (NERC CIP) reliability standards were then developed to provide a means of measuring and managing the cyber security posture of the bulk electric system. Under FERC Order 706, NERC was given responsibility for developing the CIP standards as a method of executing FERC's regulatory responsibilities. As the task of ensuring that the CIP and other reliability standards are appropriately

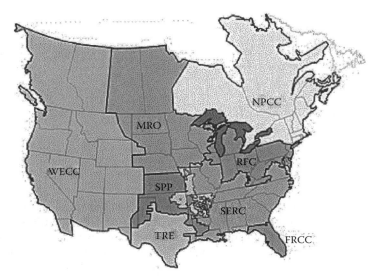

FIGURE 2.1 NERC regional entities.

executed, NERC leverages regional entities. These regional entities include investor-owned utilities; federal power agencies; rural electric cooperatives; state, municipal, and provincial utilities; independent power producers; power marketers; and end-use customers. The entities include (Figure 2.1)

- Florida Reliability Coordinating Council (FRCC)
- Midwest Reliability Organization (MRO)
- Northeast Power Coordinating Council (NPCC)
- Reliability First Corporation (RFC)
- SERC Reliability Corporation (SERC)
- Southwest Power Pool (SPP)
- Texas Reliability Entity (TRE)
- Western Electricity Coordinating Council (WECC)

The regional entities execute their responsibilities through compliance audits. NERC CIP and other reliability standards are broken down into a methodical format. You can expect the purpose to be clearly stated for each standard. Each standard also comes with an applicability clause that illustrates the responsibility entities that are on the hook for meeting the standards. The following list represents the list of responsible entities that are required to at least review the CIP 002 NERC CIP standard in order to determine whether or not they need to comply with the remainder of the standards (003-009):

- Reliability coordinator
- Balancing authority
- Interchange authority
- Transmission service provider
- Transmission owner
- Transmission operator
- Generation owner
- Generation operator
- Load-serving entity
- NERC
- Regional entity

The *reliability coordinator* is responsible for the reliable operation of the bulk electric system, has the wide area view of the bulk electric system, and has the operating tools, processes, and procedures, including the authority to prevent or mitigate emergency operating situations in both next-day analysis and real-time operations.[4] A *balancing authority* "integrates resource plans ahead of time, maintains load-interchange-generation balance within a balancing authority area, and supports interconnection frequency in real time."[5] The interchange authority "authorizes the implementation of valid and balanced Interchange schedules between balancing authority areas, and ensures communication of interchange information for reliability assessment purposes."[6] A *transmission service provider* "administers the transmission tariff and provides transmission service to transmission customers under applicable transmission service agreements."[7] A *transmission owner* "owns and maintains transmission facilities."[8] *Transmission operators* are "responsible for the reliability of their 'local' transmission system, and are responsible for directing the operations of transmission facilities."[9] A *generation owner* "owns and maintains generation facilities."[10] *Generation operators* "operate generating unit(s) and perform the functions of supplying energy and Interconnected Operations Services."[11] A *load-serving entity* "secures energy and transmission service (and related interconnected operations services) to serve the electrical demand and energy requirements of its end-use customers."[12] *NERC* and the *regional entities* are also required to review this standard.

All of these functions identified by NERC in the CIP standards represent the functions required to manage and maintain the bulk electric system. This is primarily because the bulk electric system requires a high degree of communication and coordination among different utilities. Utilities then typically do not see themselves as competitors with each other, but rather as partners in the management of the bulk electric system.

The CIP 002 standard is then used by entities that meet the criteria listed above to determine if they will need to comply with the NERC CIP standards. The requirements listed in CIP 002 illustrate the process that must be executed in order to accomplish this task. Each entity is then required to declare assets within its area of responsibility as either critical or not critical, depending on the results of the execution of its compliant methodologies. Once an asset is declared critical, the entity must determine if any cyber-assets are critical. If critical cyber-assets are identified, then the remainder of the CIP standards become applicable. Each entity with critical cyber-assets is then subjected to audits by the applicable regional entity. The regional entity utilizes the measurements found in the standards documentation to measure compliance with the standards. NERC goes further in refining what these measurements are through the development of "Reliability Standard Audit Worksheets," which are used as the checklist to determine compliance.

The manner in which NERC CIP is set up is important because it illustrates the process by which the cyber security infrastructure will be implemented for part of Smart Grid. NERC CIP is really the foundation on which utility companies are building out their security practices. This is mostly because the framework is well understood and familiar to utilities. It is important to note, however, that these standards do not apply to other parts of Smart Grid, mostly because FERC has no regulatory responsibilities as associated with the delivery of local and retail power. The compliance framework should also not be confused with the mechanism to secure the bulk electric system, but rather it is the infrastructure that will facilitate security. Because of this, many utilities will implement personnel and systems to augment compliance that will ultimately lead to cyber security practices.

In 2007, the Energy Independence and Security Act of 2007 was passed; it called for the "...optimization of grid operations and resources, with full cyber security." As a result, Smart Grid efforts were meant to ramp up. But the most important component of this legislation called for the implementation of full cyber security during grid optimization or during the implementation of Smart Grid systems and services. This law then provides the legal justification for all utility companies that wish to implement Smart Grid components to ensure that they were implementing cyber security into their Smart Grid program life cycles. The key term "full cyber security" has not been clearly defined but it is likely being interpreted through the distribution of stimulus funding under the American Recovery and Reinvestment Act (ARRA). The ARRA requires utilities that have accepted Smart Grid funding for approved projects to implement cyber security into the life cycle of their projects. While no formal regulation such as NERC CIP has been established for these efforts, the requirements are forcing utility companies to develop cyber security programs to deal with this. From a strategic perspective, it is having the effect of integrating cyber security into the Smart Grid.

As these Smart Grid projects broaden and utilities undertake efforts to implement new Smart Grid components, cyber security will more than likely be addressed at some level.

2.3 PRIVACY INFORMATION IMPACTS ON SMART GRID

Personal or privacy information is a subject generally not associated with the utility industry. But the truth is that the utility industry has been managing customer information since its inception. The management and delivery of retail power to customers requires that the utility companies maintain a capability around this. The question is whether or not there are laws, rules, or regulations that should apply to these companies in order to protect that information. Privacy information is defined in four basic dimensions, including personal information, personal privacy, behavioral privacy, and personal communications privacy.[13]

Personal information is defined as "aspects of an individual, or identifiers specific to an individual."[14] As most people know, when you set up an account with an electric company, the utility company will likely require you to complete an application, similar to what you complete when you apply for a credit card. The form usually requires, at a minimum, your name, address, telephone number, and social security number. A name alone probably does not fit the description associated with the personal information description. However, combining the name with an address likely meets these criteria because of the uniqueness of the address associated with the name of the person living at the address. Moreover, utilities must collect and verify this data so that they can be sure that the person opening the account is really the person residing at the given address. Acceptance of this information almost certainly qualifies as personal information because it is meant, by definition, to denote that a person's identity is correct. The social security number can further be used to provide authentication that the data gathered clearly represents that of the applicant.

Personal privacy information is defined as the rights to control the integrity of one's body. The information collected by a utility likely does not meet these criteria because the utility likely cannot use the information to control anything regarding the integrity of the applicant's body. Personal privacy information likely does not represent information that a utility wishes to have, nor is it data that is needed.

Behavioral privacy represents the "right of individual to make their own choices about what they do and to keep certain personal behaviors from being shared with others."[15] While information associated with an application does not alone constitute anything regarding behavioral privacy, but combined with energy usage it may. The result is that utility companies may be concerned about behavioral privacy with the implementation of Smart Grid, combined with customer energy usage data. Smart Grid at the consumer level is about trying to manipulate users' behavior by providing them incentives to change behavior. This can be illustrated through time-of-use type programs, which likely result in a utility's ability to profile users and encourage them to change their behavior based on incentives. It should be pointed out, however, that the change in behavior is totally up to the consumer. However, if a threat were to gain access to this information, they could understand how a person uses his energy and potentially know when a consumer is at home or not. This information could of course be used by adversaries to potentially conduct crimes against the utility's customer.

Personal communications privacy refers to the "right to communicate without undue surveillance, monitoring or censorship"[16] For this to be an issue in the Smart Grid space, utilities would have to use the information they acquire from energy usage to monitor the consumers' communications. As this is not the goal of Smart Grid efforts, it is likely that it does not represent something about which utilities should be generally concerned.

Based on these dimensions, utilities have a general concern associated with privacy with regard to personal information that they knowingly collect and the potential for behavioral privacy. This is all related to what could happen if a cyber security threat were successful in stealing this information. As a result, utilities may be required to take action in an effort to prevent such data theft in order to defend themselves against any liabilities. This is mostly because they have to collect personal information in order to conduct business and will have to collect information that will lead to a breach in behavioral privacy as Smart Grid programs are implemented.

Your name, address, and telephone number are required so that the utility can catalog you as a customer and ensure that you reside within their service territory. This information may be considered personal or private, but it is relatively routine that consumers would provide this information. For example, when you acquire a telephone number, the phone company gathers this information and then publishes it in the "white pages." While you have the option to pay to have your number listed as private, the utility companies are not in the business of publishing your information. They do, however, keep this information and this is generally accepted.

Social security numbers, however, are also required and used to conduct a financial history check in order to determine the applicant's ability to make monthly payments on their utility bill. The financial history check likely includes the execution of a credit history check, which means that the utility companies must acquire one's credit score. The credit score and social security number must then be protected because they do represent personal and private information. The storage of this information means that utilities are responsible for protecting this information. This information is a likely target for attackers who wish to gather this data and then use it for potentially malicious purposes.

To deal with these potential effects, it is likely that utilities will need to execute privacy impact assessments in order to determine if they are really capturing privacy information. And if privacy information is captured, then they will have to protect it in alignment with potential laws that might apply, including but not limited to the Fourth and Fourteenth Amendments to the United States Constitution, the Privacy Act of 1974, and other laws. This of course all hinges on the likelihood that threats will break in, steal, and use this information for malicious purposes. If a social security number is stolen, for example, and this information is used to commit fraud, then the utility company would likely be found liable to the consumer for allowing the fraud to occur on their computer systems. The likely impact would be that the utility company would be required to pay for all damages associated with the theft and fraud, including payback for any money lost, restitution for the hardship faced during the loss, and, finally, identity protection for life. For one consumer, this may not be challenging, but considering that these systems may potentially store information for millions of customers, it could be extremely detrimental to the utility.

If a threat were to break into a utility that maintained a customer base of 3 million and simply steal every single social security number, what would the impact be? Considering that identity protection costs somewhere between $5 and $15 per month, taking the low cost as an example, the monetary impact to the utility losing that much information may result in a financial liability of $5 multiplied by 3 million customers. This would then represent a cost to the utility company of $15,000,000 per month every month for the life of every customer whose data was stolen. This could then represent a liability of $180,000,000 per year, which is a significant cost. As reaction to something potentially occurring, a utility might expect to employ forensics experts to come in and disprove the fact that the information was stolen or to simply try to pacify their customer base. The bottom line is that their consumers will lose trust in the company's ability to protect their information and this could potentially have a significant impact on the overall business model. It pays then for the utility company to make an investment in cyber security in order to at least prevent an event like this from occurring. The key will be that they would have to show that they were not negligent in storing this information and this is where cyber security comes into play.

The stronger the cyber security posture of a utility, the more it will be able to prove that it was not negligent in the storage of such information. The more it can prove this, the easier it will be for them (the utility) to shift the burden of loss back to the consumer or taxpayer. A business action that might be taken is to buy insurance and if this ever begins, we might start seeing the insurance companies to start actually regulating the utilities for the storage of privacy information. Some action will need to be taken as systems become more interconnected through the Smart Grid in order to prevent potential loss of privacy information to malicious threat sources.

2.4 SECURITY STANDARDS

Compliance has played a major role in *information security* over the past decade. Although there have been various organizations such as the International Standards Organization (ISO), Information Systems Audit and Control Association (ISACA), etc., none have achieved more attention than the National Institute for Standards and Technology (NIST). The E-Government Act (Public Law 107-347) passed by the 107th U.S. Congress and signed into law by the President in December 2002 recognized the importance of information security to the economic and national security interests of the United States. Title III of the E-Government Act, entitled the Federal Information Security Management Act (FISMA), requires each federal agency to develop, document, and implement an agencywide program to provide information security for the information and information systems that support the operations and assets of the agency, including those provided or managed by another agency, contractor, or other source. One of the major elements of this law included in H.R. 2458-52 states the following:

> "'(3) delegate to the agency Chief Information Officer established under section 3506 (or comparable official in an agency not covered by such section) the authority to ensure compliance with the requirements imposed on the agency under this subchapter, including—
> '(A) designating a senior agency information security officer who shall—
> (iv) head an office with the mission and resources to assist in ensuring agency compliance with this section;'
> and
> '(B) developing and maintaining an agency wide information security program as required by subsection (b).'"

Additionally, under § 3544 identified Federal Agency Responsibilities, which include the fact that they are required to develop and implement an agencywide information security plan to provide information security for the operations and assets of the agency. The head of each agency is responsible for adequately ensuring the confidentiality, integrity, and availability of information and information systems supporting the agency operations and assets; developing and implementing information security policies, procedures, and control techniques sufficient to afford security protections commensurate with the risk and magnitude of the harm resulting from unauthorized disclosure, disruption, modification, or destruction of information collected or maintained by or for the agency; and ensuring that the agency's information security plan is practiced throughout the life cycle of each agency system.

The FISMA Act of 2002 was really the first major attempt to establish the field of information security formally outside the Department of Defense (DOD) and the National Security Agency (NSA). The DOD and NSA have maintained strong cyber security programs for some time, as a result of their line of business. Virtual attacks on computer information systems are generally nothing new. The DOD, for example, is involved in warfare and information security was likely implemented long ago in order to protect their classified and confidential information from their adversaries. The same is true for the NSA and likely the Department of Energy (DOE). The introduction of FISMA basically extended those practices into the non-DOD/NSA/DOE agencies. For example, the U.S. Department of Interior (DOI), Department of Agriculture, etc. were now required to establish formal information security programs. As the need for these programs gained traction through Inspector General assessments, etc., more of these agencies began to turn to the U.S. Department of Commerce's NIST organization.

NIST Special Publication 800-26 was an attempt to get organizations to conduct evaluations of cyber security posture and to establish their level of maturity based on what was found. This was an attempt to quantify information security risks and measure the effectiveness of the organization. While this approach represented a solid approach to identifying potential weaknesses, it did not provide a roadmap for how risk could be mitigated should weaknesses be identified. NIST later came up with SP 800-53, which represented a controls-based approach to identifying risk. The SP 800-53 approach was also one of the first standards defined to be cyber security standards. The U.S. Department of Energy Power Marketing Administrations (PMAs) are federal transmission organizations with multiple control center systems throughout the United States. The Bonneville, Western, and Southwestern Power Administrations represent utility companies that are federally owned and conduct energy management on a massive scale.

Because they are federal agencies, they are required to meet the NIST-based approach to information security. This means that they had to define their FISMA systems in alignment with the Title III of the E-Government Act of 2002. The results were that their control center systems are considered "FISMA Systems" and were thus subject to the controls defined in NIST SP 800-53. Meanwhile, FERC issued Order 706, which required the development and implementation of the NERC CIP Standards, for application to North American utility companies. As the PMA federal agencies are also utilities bound to the oversight of NERC, they were required to also meet the NERC CIP standards, in addition to NIST SP 800-53 at the same time. This was also true for other federal entities that provided power operations and generation support for the general public such as the Bureau of Reclamation and Army Corp of Engineers who many times own and operate the major dams throughout the United States. This may have, in effect, been the genesis of where the need for greater Smart Grid security standards actually began. The PMAs, BOR (Bureau of Reclamation), and the Army Corp of Engineers are unique in this light because federal oversight is conducted by the Inspector General, who has responsibility for oversight of their control center systems with regard to cyber security posture. At the same time, NERC was prepared to regulate those same control center systems with its CIP standards. When analyzing the impact of implementation between the two sets of standards, it becomes inherently obvious that one set understands the concept of reliability while the other pushes cyber security in a manner that is a risk to reliability.

The NERC CIP standards are process oriented, as they emphasize the need for procedures versus the actual implementation of security controls on the assets themselves. Although they bring the concepts of cyber security to the control center environment, they do it in a manner that will have the least impact on reliability. The concept of asset management according to NERC CIP is to figure out what the most critical cyber-assets are that are needed to manage bulk energy and effectively exclude all others from regulation. Once the cyber-assets that need to be protected are identified, NERC CIP focuses on the need to develop an access control program capability in order to manage access to the environments where they reside in addition to other security control processes and procedures. This is done through multiple standards within the NERC CIP framework in order to control physical access, electronic access, access to electronic and physical information, as well as access to badge readers and firewalls. The standards are unique because they require the development of a process to ensure that personnel with any of these access types complete a background investigation and have cyber security training related to their organizational role, and that all specific access types, including records, be maintained and available for audits. Because the audits are measured, utilities are forced into a situation where they must clearly manage and maintain these records, resulting in the creation of effective processes. Security controls on assets are addressed on a limited scale, where the standard requires the implementation of patches, anti-virus, log monitoring, account management, vulnerability assessments, procedures to test cyber security posture when changes are made, as well as the need to dispose of assets properly prior to removal. And finally, disaster recovery and incident response processes must be developed, implemented, and tested in order to meet the standards. NERC also out-lines a process for taking exceptions to the standards, which are executed whenever the utility comes across a system that cannot implement the security controls that have been prescribed. The benefit of implementing these standards in this fashion is that it is really left up to the utility on how to do something, and it is known that whatever they decide to do, it must produce the records required for an audit and these records are mea-surable. The downside becomes however that overall cyber security risk is not really addressed as the utilities instead focus on complying rather than understanding the threat.

NIST takes on a different approach to securing the architecture. The E-Government Act of 2002 requires that whole architectures be identi-fied fully, and that all security controls and the concept of risk manage-ment be applied to those architectures. Furthermore, the list of NIST SP 800-53 controls is much more detailed and much more expansive. For example, there are eighteen control families, with multiple controls spread out through each family, all segmented by the concept of impact that has been established for the architecture in alignment with NIST SP 800-60, "Guide for Mapping Types of Information and Information Systems to Security Categories." The approach requires that all your sys-tems be identified and each one quantified for impact. Based on impact, you are required to apply controls defined in NIST SP 800-53 using the high, medium, and low concept as represented in the impact study. Furthermore, there is no process for taking exceptions, only the abil-ity to identify and accept risk. NIST SP 800-53 emphasizes risk man-agement while NERC CIP focuses on compliance. Both are intended to elicit security: one achieves that through the identification of risk and its acceptance with a broad scope, while the other achieves success through scope limitation and the requirement to achieve specific measurable objectives, including reports and audit trails that prove implementation. Figure 2.2 illustrates the key differences between the two concepts.

Although it is not certain how NIST came to focus on energy man-agement systems, it is clear that the PMAs, BOR, and the Army Corps of Engineers had some influence on this because of the need to satisfy both NIST and NERC. Needless to say, NIST began to include secu-rity controls focused on transmission organizations into their 800-53

NIST SP 800-53	NERC CIP
All systems are defined	Systems defined by function
Criticality used to identify impact level	Critical Impact is the only level identified
Risk Management approach	Compliance-based approach
Confidentiality, Integrity, Availability	Reliability

FIGURE 2.2   NIST versus NERC.

standards. This was done even while NIST understood that NERC CIP was the regulatory authority over transmission control systems. NERC has since begun the development of version 4 of its standards. This version proposes an impact-based concept that should include the full scope of energy management. The concept of energy management represents multiple facets within a utility company. These facets are usually represented as Generation, Transmission, Distribution, and Metering. Each of these facets plays a role in the management of energy throughout the United States, and various utilities play different roles in each. Furthermore, NERC, understanding that a more comprehensive solution for securing cyber-assets is coming, is proposing to allow for other cyber security frameworks such as ISO, NIST, COBIT (Control Objectives for Information and related Technology), etc. to be used to manage cyber security. While these frameworks may not be effective in the management of control system security, the flexibility is meant to lead organizations to draw their own conclusions or potentially develop a more robust energy-sector framework.

This leads us to today, where NIST has drafted NISTIR 7628, "Smart Grid Cyber Security Strategy and Requirements." This document is the latest in a line of planning elements that were initiated in order to develop a strategy for securing both energy management and Smart Grid systems. The term "Smart Grid" is more and more becoming the terminology of choice when identifying energy management systems, whether they are in metering, distribution, generation, or transmission systems. The point is that appropriate planning is going on to develop a more comprehensive solution that takes the best parts of NIST SP 800-53 and aligns them with the concept of reliability and the objectives that NERC wishes to achieve.

2.5 SMART GRID SECURITY STRATEGY

Vulnerabilities are identified based on gaps in the required security controls. That is, you need to know which security controls are needed in order to identify risks using internal and external threats as tools that threaten the system in question. There are security controls that prevent external threats from breaking into a system and controls to augment security associated with internal threats. In both cases, there is always the threat that someone or something will attempt to access your environment, compromise your systems, and create a liability for your organization. The liability to the organization represents the impact in the risk equation probability x impact = risk. To measure risk, you need to understand the impact and what the likelihood is that the impact will be realized. This is an important concept to grasp when it comes to the world of Smart Grid because the traditional boundaries that segmented external from internal threats are no longer relevant. Furthermore, the impacts that can be realized in Smart Grid extend beyond the world of identity theft and into the realm of risk to human life and even humanity itself.

Threats are generally constant because they threaten the organization's ability to secure itself against risks that have been identified. The difference between an external and an internal threat has to do with the concept of cyber-risk versus cyber security risk. A cyber-risk is defined as a risk that threatens the reliability of a cyber-asset within a Smart Grid architecture. This includes the potential that a staff member may improperly operate a Smart Grid system, resulting in an undesired impact. An example could be an employee tripping over a power cord in the data center and causing a server that supports Smart Grid technology to stop working. This is a cyber-risk because it is not a risk as a result of a malicious threat, while a cyber security risk should be associated with a malicious event that threatens to cause the same impact or liability to the organization. For example, if an employee *intentionally* tripped over a power cord, then this is a cyber security risk because that person intentionally tried to cause harm to the stability of the system. The two concepts are important to understand because of the way that utilities view risk to their energy management platforms. The goal of the cyber security professional should be to identify whether or not a cyber-risk is actually a cyber security risk. And this is the problem represented when discussing how to secure Smart Grid technology. A malicious external threat source has the intention of creating the same cyber-risks that utilities are concerned about; however, the view from the utilities is that cyber security threats are somehow different from cyber-threats. This difference has to do with external versus internal threat. In other words, utilities are concerned with internal threats and are happy to implement controls to prevent them, because there is a long history of them causing outages and events that impact the reliability of their energy management systems. Whereas, external threats represent traditional hacker-type risks because there is a limited history of them impacting their energy management systems.

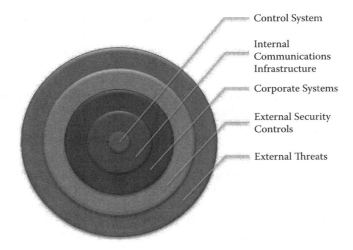

Control System

Internal
Communications
Infrastructure

Corporate Systems

External Security
Controls

External Threats

FIGURE 2.3   Control system location.

Whether it's internal cyber-risk or external cyber security risk, controls must be implemented to mitigate both risks types. The general method of securing against these threats is to implement technical controls, usually network based, that prevent external threats from accessing or even seeing the energy management system, for example. Most utilities have isolated their control systems, which end up being networks buried deep in their corporate networks. There are almost always firewalls between networks; that is where the air gapped view derives. The risk that is quickly becoming an issue is where these access control points are opened so that the operational users can access their equipment more easily. Going back to the example of remote access to relays, it will be much easier to have all the technicians essentially sit at their desks and control them from their corporate desktops. From a reliability perspective, this is good because the quicker they can get to them, the faster problems can be solved. However, from a cyber security perspective, even with the use of VPNs, point-to-point firewall rules, etc., the door is opening from the corporate network into the operations network. This implies that the air gap will slowly go away over time and threats may start to find their way into the control systems. Figure 2.3 illustrates where the control systems reside in relation to the common network configuration.

External threats generally threaten the external security of the organization. As the control system is protected at the core of the organization, it are perceived as not vulnerable (despite vulnerabilities that exist in their systems) to external threats. Keep in mind what was stated earlier, that traditional IT security controls many times cannot be applied to these assets, so if they were ever exposed to traditional IT threats such as malware, it would not take much to knock them over. Moreover, there is merit associated with this utility belief because there have not been any or many reported penetrations of networks all the way into the control system that have resulted in the big impacts with which utilities are concerned. Instead, utilities are more concerned with internal threats, of which there is a lot of evidence that demonstrates that large impacts have been realized as a result of cyber-risk. These risks are generally associated with operator error or conditions in which a control was used that should not have been used based on a miscalculated condition.

Internal threats are secured against using process control, hence "process control system." Control systems are effectively just processes that are used to control other processes. So the natural method of securing process control systems is through process control itself. This is why the NERC CIP standards are written from a process-oriented perspective. One should track and manage users who have access to the environment so that they know that access is privileged, and they understand what their responsibilities are associated with such access. This does not mean that access is restricted for users; it just means that their access is known and hopefully the risk of such access will be identified at some point in the future. Whereas, the NIST-based approach takes into account the risk and is looking for more of a role-based access control solution. This is not to say that NIST does not take this into account; it simply illustrates the point that NIST views the problem differently from NERC. Remember that NERC is really about allowing everybody to continue what they are doing so that reliability is not impacted, but the utility should just know who has access. NIST is more about looking at who has access and trying to figure out whether or not they really need it. Both standards have their flaws, NERC from a cyber security risk perspective and NIST from a reliability perspective. The reason that utilities prefers the NERC approach is that they believe and trust in their trained engineers and they know that their number-one priority is reliability, so they need to be able to do whatever they need to do to make the system work. If an engineer ever gets into a situation where she cannot do something that she needs to do because she was viewed as not needing the access to be able to ever do it, we could potentially see a power outage that was caused because access was restricted—that is the fear. Keep in mind that these are real-time systems, and reaction to normal events may only allow for a response to occur in seconds; if that response does not occur, then outages and problems become possible.

The way this should probably be approached is that controls should be categorized based on impact to the control system, and more flexibility should be allowed where those controls have a higher propensity to cause reliability concerns. At the same time, the capability to react

Req ID	Control Name	Gap Description
AC-1	Access Control Policy and Procedures	No policy exists which requires that access controls be in place.
AC-2	Account Management	There is no common process for managing user accounts for assets within the architecture.
AC-3	Access Enforcement	There is no mechanism which enforces access controls to and from assets within the architecture.
AC-4	Information Flow Enforcement	There is no mechanism to control the flow of information to and from assets within the architecture.
AC-5	Separation of Duties	Role-based access control is not implemented for assets within the architecture.

FIGURE 2.4   Security control gaps.

to security through monitoring needs to advance to a point where it can be used to prevent and respond to cyber security-related events. For example, if an engineer does something that is out of the ordinary, and his action is malicious, then cyber security should understand what that is and how to react to it. The risk of preventing something is probably greater than the risk of a cyber security event happening, so a reactive approach will likely be the best option. Weighting of controls based on the mission of the organization then must be taken into account in order to integrate cyber security standards that will be viewed as actually augmenting security posture.

The point is that when evaluating risk to a Smart Grid architecture, one needs to understand that there will always be risk based on internal and external threats. Second, a security controls approach should be used to identify gaps so that they can be analyzed for risk. Using the NIST SP 800-53 set of security access security controls, if architecture has gaps associated with each of the controls represented in the following figure, then what are the risks? See Figure 2.4.

The risks associated with these gaps must be identified in the context of external and internal threats. Given the gap, what weakness is present, and can that weakness be exploited by an internal or external threat? Figure 2.5 breaks down each gap according to the threats identified.

In this case, a business case has been made for implementing a control to mitigate the risk of either an internal or external threat. The business case is made by illustrating the risks associated with the lack of security controls illustrated. If the organization, for example, does not have an access control policy, then why would administrators implement access control? And if administrators do not implement access control, then what will prevent internal users with no need for access to critical resources from accessing critical resources? A policy will be very effective in preventing a nonmalicious act by a user because the user would be aware of something he is not supposed to do, but at the same time not actually restricted from doing what he needs to do in order to react to something that needs to be done. This would even be effective where the security control cannot be implemented to prevent access. The external threat, however, is not going to care about the policy as it is written and with no access controls in place will simply access the resources that he knows he is not supposed to access. Based on the example above, we know what the risk is and we know that the implementation of the control will prevent the risk to some extent. However, what happens if the control cannot be implemented because the control itself represents a risk?

As is the case in Smart Grid architectures, security controls themselves represent liabilities, more so to control systems. Using the example from Figure 2.4, it is obvious that policy can be implemented, but a technical solution to actually restrict access may not be possible because it represents another hop for the assets and thus could result in bad performance. As bad performance is counter to the support of reliability, such a control may represent a cyber-risk to the architecture itself. So if the control represents a risk and lack of the control represents a risk, then how is risk mitigated? The answer is through the identification of an alternative control. This will more than likely be a process control that places a restriction on behavior so that a technical control can be implemented to detect anomalies in behavior. Again referring back to our list of gaps, where controls cannot be implemented, what alternative controls can be used to mitigate the risks identified? Figure 2.6 illustrates this.

The key to securing Smart Grid architectures using risk-based techniques is then to identify the mitigation strategy for a risk that exists because a security control is not present and associate it with what an internal resource might end up doing and the impact that would be caused as a result. Identifying the cyber-risk is likely more important than the cyber security risks because they are more likely to occur. Moreover, utilities will pay more attention to cyber-risk because they understand that the chance of occurrence remains high without mitigation because accidents have occurred in the past. What should be noted, however, is that if an external threat were to gain access to internal resources, that threat would attempt to maliciously cause accidents to occur that would ordinarily be caused by an internal threat. Understanding this concept will help utilities and security professionals understand that the reality is that there is no difference between an impact that can be caused by an internal or external threat. Therefore, the security controls should be implemented

Control Name	Internal Threat	External Threat
Access Control Policy and Procedures	Without access control policies and procedures users may access systems for which they are not supposed to and are not trained to use, resulting in the potential that the systems may be used in an unauthorized manner.	Lack of an access control policy and procedure may lead to not properly implementing controls which prevent external threats from accessing internal resources.
Account Management	Without account management there is no way to differentiate between users who are supposed to access systems and those who are not.	Lack of account management may make it easier for an external threat to escalate privileges on user accounts if they have penetrated the architecture.
Access Enforcement	Without access enforcement there is no way to restrict access to systems by users who should not have access.	Lack of access enforcement may allow for an external threat to access the architecture from the outside because access restrictions may not be implemented.
Information Flow Enforcement	Without information flow enforcement there is no way to control what users do with information from systems within the architecture.	Lack of information flow enforcement may lead to an external threat gaining insider information about the makeup of the target architecture.
Separation of Duties	Without roles-based access controls there will be no way to identify who caused a cyber-risk in the event that one occurs because all users will have equal access despite their access needs.	Lack of separation of duties may allow for an external threat which has compromised a user account to access any resource inside that they desire.

FIGURE 2.5   Threat assessment.

in order to prevent both threats from being able to successfully create an unwanted impact. Moreover, cyber security professionals are going to be able to mitigate far more risk from the perspective of an internal threat than they will by simply focusing on an external threat. And they will, at the same time, be preventing an external threat from exploiting internal weaknesses should access control barriers ever break down.

Control Name	Process Mitigation	Technical Mitigation	Threat Mitigation
Access Control Policy and Procedures	Develop and implement an Access Control Policy and Procedures for provisioning access.	Centralize management of access control policies, so that users are forced to engage an authority to be provided with access.	Internal Threat contained by process mitigation. External Threat contained by technical mitigation.
Account Management	Implement an account provisioning process.	Centralized account management solution that reports changes to user accounts.	Internal Threat contained by process mitigation. External Threat contained by technical mitigation.
Access Enforcement	Require all access be default deny and require a change control process to open needed access.	Reports of changes to access controls.	Internal Threat contained by process mitigation. External Threat contained by technical mitigation.
Information Flow Enforcement	Implement an information characterization and classification process.	Data loss and prevention solution.	Internal Threat contained by process mitigation. External Threat contained by technical mitigation.
Separation of Duties	Develop roles and associate users to role types.	Implement roles-based access controls associated with account management.	Internal Threat contained by process mitigation. External Threat contained by technical mitigation.

FIGURE 2.6   Risk mitigation.

2.6 SMART GRID IMPACTS

NERC CIP 002 Critical Asset Identification illustrates a methodology for identifying assets that are critical to the bulk electric system. This methodology, however, does not truly address all aspects of utility assets, including distribution, advanced metering, and some generation assets. This is an important notation because assets that make up Smart Grid must be evaluated for impact before we can understand the risks associated with the cyber-assets within them. The current NERC CIP methodology takes into account assets that meet the following conditions:

- Control centers and backup control centers
- Special protection schemes
- Automatic load shedding
- Generation resources critical to the reliability of the bulk electric system
- Transmission resources critical to the reliability of the bulk electric system
- Assets critical to system restoration

The most recent version of this methodology, known as CIP 002 version 4, further defines how these criteria end up being critical by applying the bright line concept. This bright line concept is only applicable to the transmission and generation resource requirements, providing clear guidance on what is and what is not critical. For example, generation facilities that have a total output of less than 1,500 megawatts are not considered "critical assets." This methodology has been established to address the criticality of assets for the bulk electric system. While this is important, it is equally important to understand the impact for the other areas of Smart Grid because risk cannot truly be quantified until impact has been determined.

NIST SP 800-60, "Guide for Mapping Types of Information and Information Systems to Security Categories," perhaps provides a framework for evaluating the criticality of assets within Smart Grid. While the methodology does not necessarily apply to Smart Grid, it can potentially be modified to further provide a way to measure impact for Smart Grid assets. The point is that a library of impacts for each component of Smart Grid must be developed so that it can be used to evaluate the risk to confidentiality, integrity, and availability. This can be accomplished when viewing each component within Smart Grid by asking simple questions in relation to confidentiality, integrity, and availability. Figure 2.7 illustrates a method for simplistically identifying impact. Security Area references the standard confidentiality, integrity, and availability associated with the impact type. *Confidentiality* refers to the impact of disclosure of something associated with the asset type. The loss of information associated with the asset, for example, and whether or not the impact of such an event is high, medium, or low based on the view of the organization. *Integrity* refers to a loss of integrity, which is more closely associated with the concern that the asset may be controlled by

Security Area	Impact Description	Impact Discriminator	Impact Risk
Confidentiality	Loss of Information	What is the impact to the system if information is lost or stolen?	High, Medium, or Low
Integrity	Loss of Control	What is the impact to the system if it is hijacked by an unauthorized party?	High, Medium, or Low
	Loss of Data Integrity	What is the impact if the system provides invalid data?	High, Medium, or Low
Availability	Short-Term Denial of Service	What is the impact if the system is down for less than 48 hours?	High, Medium, or Low
	Long-Term Denial of Service	What is the impact if the system is down for more than 48 hours?	High, Medium, or Low

FIGURE 2.7   Simple impact table.

an unauthorized party. And *Availability* deals with the concern if the system were made to be unavailable.

An example of how the table in Figure 2.7 might be used to determine impact on a Smart Grid asset would include an analysis of this information associated with the asset. If, for example, we were talking about a smart meter communications infrastructure, how would these questions be answered? What is the impact if the system (asset) were to be hijacked by an unauthorized party? The reviewer would ask questions along these lines to view part of the impact of having an integrity breach in the context of cyber security. What could an attacker do, for instance, if she were able to hijack the communications infrastructure for a smart metering deployment? One answer might be that she would be able to monitor all communications and potentially redirect where that information goes to. The result could be that the utility that owns the deployment may not get the data it needs to effectively bill the customer. And the question to the utility should be: how would this affect our business? The utility will most likely be concerned with the likelihood that something like this could occur and should it, the impact would likely be high for a number of reasons. It is important to understand those reasons, which can later be used to demonstrate why a certain control should or should not be implemented.

Impact goes hand in hand with the application of security controls. This is because the lack of controls leads to a heightened probability that the unwanted event can occur. There are various models for impact

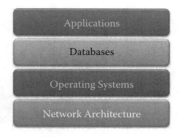

FIGURE 2.8   IT-enabling asset types.

categorization as discussed above, and those should be used to improve the ability to identify impact. The key is going to be making it as simple as possible so that the organization can easily recognize that something can happen and that mitigation of the impact should be of primary concern. Creating a list of impacts associated with different scenarios also allows for the identification of security project tasks that can be used to mitigate the risk. Tying these directly back to a control standard will likely justify why it is being implemented. The more that a security control can mitigate an impact on one or more assets, the more likely it is that the implementation of that control will be accepted.

2.7 APPLYING SECURITY CONTROL FRAMEWORKS TO SMART GRID

There are four common sets of assets regardless of architecture type: applications, databases, operating platforms, and network devices. Smart Grid is no different; however, it is perceived to be because embedded versions of operating systems are used or industrial control systems that function based on logic are used. Regardless of this, they still represent the four basic forms of information technology-enabling assets (Figure 2.8).

Industrial control components, meters, and other devices that make up Smart Grid but are not applications, databases, operating systems, or network architecture are at least a hybrid of several or rely on control from an application. In each of these cases, these things must be secured independently of process in order to mitigate risk to the components.

There are a few basic problems with today's risk management models, in the context of 800-53. NIST SP 800-39 provides guidance on risk management but assumes that you have the resources to properly execute and acquire all that information. NIST SP 800-53 maintains more than 500 controls that depend on how the system being analyzed is classified. Furthermore, it does not really take into account the specific components of the system being analyzed. And then there is the question of

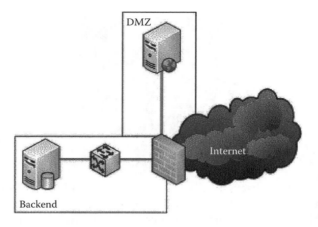

FIGURE 2.9   Example generic network architecture.

what constitutes a system or a network. Defining *system* literally in broad terms is probably the most accepted method and then to categorize *assets* based on what they are used for in supporting the system. For example, a *Web portal* may be considered a "system," whereas the Apache Web server, Linux operating platform supporting Apache, the Oracle database that the Apache Web server connects to, and the Microsoft Windows server that supports the database should all be considered assets.

The web server application is an asset because security controls can be implemented on it, for the same reason that the database and operating systems should be considered assets. One might even consider the network components that manage connectivity for these assets to be assets within the same system themselves. The general view of such an architecture appears in Figure 2.9.

The "system" in this case would be the entire architecture illustrated. The assets involved are less obvious; the web server is really two assets, while the switch and the firewall represent two assets and the database server and associated operating system represent two assets. In total there are six assets, all of which have different security settings. From a risk management perspective, the question is whether or not you want to have a switch brokering access to your database server, which has limited security capability implemented, or do you, for example, want to restrict what is allowed to communicate with that database server at the switch and firewall? Then there is the larger question of how all of this will be managed. NIST provides guidance on how to conduct the system assessment for Technical, Operation, and Management Controls; the problem is that most people conducting such an assessment really won't understand that access control applies to all six assets in this case.

You can develop policies that mandate that these systems meet certain criteria, or you can start from the bottom and try to harden all the assets. The trouble is that you never truly accomplish your goal, and it is still difficult to measure risk in this instance. There are numerous combinations that you can use to determine what the risks are, but quantitatively it is difficult—if not impossible—to do. And then when you begin thinking about how many assets there are and how many systems an entity might have, it is difficult to imagine running through the full NIST SP 800-53 control catalog for every system; especially considering the costs, does it really outweigh the benefit? In either case, the goal is to understand which security controls are implemented from an operational management perspective for the overall system and what the technical security posture of each individual asset is.

Assessing each asset independently using the technical controls from within the technical controls family is probably the quickest method for building a unique asset security profile. You can ask simple questions such as does the web server require all users (administrators and end users) to identify themselves? If the answer is yes, then the next question relates to what kind of authentication is required and what is the extent of that authentication. If the answer ultimately ends up being that all six of my assets require unique identification, and they all require eight-character passwords for use and token-based authentication for administration, then I know that my asset risk is relatively low in terms of this control. If I, however, do not have any logging mechanism set up and I do not lock out accounts after unsuccessful log-on attempts, never apply patches, etc., then that asset risk increases significantly. This is why there are certain questions that you want to know of for each unique asset type in terms of configuration.

At the same time, you want to make sure that the organization has an operational plan for implementing operational controls across multiple "systems." These controls only represent that the entity wants certain controls implemented, but do not illustrate whether or not this is actually true. For example, I might write a plan on how I am going to control change in my environment, aligning with the NIST SP 800-53 Configuration Management family; but even with this policy implemented, there is no assurance that all of my assets follow it. Assuming I was not following the policy written, but the policy or plan was written to the exact specifications of NIST, I might end up with low risk in regard to what security management desires and high risk with what is actually implemented. The question becomes: Which risk has more weight, and how would the two be combined? And there is even a larger problem in determining the probability of occurrence because there really is no way to illustrate how threat sources might exploit your lack of controls. In other words, there are a lot of factors to consider when measuring risk in regard to information security.

Perfection in information security is generally not possible, because everything depends on something else. This is true for risk management; if I call something a high risk, then there are likely a number of reasons why it is not a high risk. To quantify, this will require a lot of work and, in reality, I am just wasting a lot of time trying to prove that something is or is not true. The reality of the situation is that the head of any organization wants to know what the risk of an information security problem is for the entire network. Whereas, their subordinates may want to know what the risk is for each function and their subordinates of the risks to the systems and so on. Based on this information, the concept of identifying weaknesses in the organization's security posture should likely be applied at the general level. So if I have no patch management plan, patch all my windows servers, and sometimes patch my Unix servers in the same "system," I might say that my system integrity risk is moderate-high. This is mostly because there is no management commitment to patch management and only one of the two unique systems types in the environment is doing what is required.

The most feasible solution is to broadly look at the controls catalog for each unique asset type and then determine what the risk level is for operational management of that environment. Figure 2.10 illustrates an example of the results needed to begin illustrating risk, using the content described above.

This approach allows you to visualize where assets are in terms of security controls and demonstrate how compensating controls can be used to protect assets that do not maintain the proper controls. For example, the firewall maintains the ability to restrict access to itself and other systems. However, the network switch may not maintain this

Operational Assessment			
Control	Description	Res	Risk
System Integrity Protections	Are patch management processes available for all platforms and applications within the environment?	No	
Configuration Management	Is there a plan to baseline the security configuration for each asset?	No	
Contingency Planning	Has a plan been developed to restore the system in the event of failure?	No	
Incident Management	Is there an incident management plan for the environment?	No	

FIGURE 2.10　Security assessment.

capability. If the firewall is connected to the switch, then you are able to specifically control who is able to directly access the switch. This concept is known as controls inheritance and is important to understand, especially in environments where controls cannot be implemented. The basic premise here is that now reds can be turned to yellow based on the ability to inherit controls. The following bullet list demonstrates how this control inheritance would be implemented using the assessment results from above:

- Apache Web server inherits access control and audit/accountability from the Linux operating system.
- Apache Web server inherits access control and audit/accountability from the firewall.
- Network switch inherits access control and audit/accountability from the firewall.
- Oracle database inherits access control, identification, and audit/accountability from the Microsoft operating system.

The controls inherited by each asset with gaps demonstrate that each asset type is still protected even though it cannot or does not meet the standard on its own. Looking at the example provided, one notices that there are inconsistent security configurations on the assets within the architecture. This is most likely because there is no plan for baselining the configuration of each unique asset type as seen in the "Operational Assessment." Furthermore, in this example environment, no plans have been developed to augment security, which is likely why there are inconsistent controls being applied. For example, the network switch has the capability to do access control, but it is not implemented because the team managing the device may not be required to turn on that feature. This is generally where you find vulnerabilities in most enterprises, because there are inconsistent controls applied and that means that gaps can exist. Threat sources (hackers or other malicious users) are generally looking for a breakdown in security that can be exploited. More advanced threats look for weaknesses in the vendor-created products, which is a separate risk that the organization cannot really control.

The controls inheritance process demonstrates that the system is really protected, but the lack of planning may demonstrate that in the changing future, this may not be the case. For example, if no plan exists for baselining the security configuration of assets before being placed into service, it may be possible that an asset replacement being done under new management will not be secured as the old one was. This is because the previous manager took security seriously and verbally required that all assets be configured securely before being placed into production. But the new manager may be more concerned about getting things done quickly and decide to not require secure configurations, thus opening the door to risk. However, if a plan or process were developed that was universally mandated, then it would not matter who the manager was, as there would be something for reference and following the plan would likely be implied.

Risk management is a concept that needs to be ongoing and used to thoroughly evaluate cyber- and cyber security risk in order to identify weaknesses and their planned remediation. To properly understand likelihood, we need to understand the risks associated with the individual cyber-assets. We then need to pair that risk with the overall impact identified with each of the assets where the individual cyber-assets are located.

2.8 MANAGING THE OVERALL RISK TO SMART GRID

Understanding the utilities and how they relate to one another including the relationships with vendors and the government at the federal, state and local levels is key to a good risk management strategy. Utilities as most companies, are typically vertically oriented, as they maintain business areas that support the various areas of grid management. Understanding these verticals and why they exist will allow for the development of a risk management strategy. Verticals within a utility are defined generally to serve a specific function. For example, there are organizations within grid operations that support the various elements of the grid. Substations are usually managed by a group that is separate from the group that maintains the control center. And power generation groups are usually independent of substation and control center groups, as well as autonomous amongst themselves. Finally smart meter groups are different from all three of the others, as they are focused on power delivery and billing at the consumer level. The separation of duties between these groups and the power plants contributes to the complexity in understanding the cyber risk associated with each. This is because each group views the power grid differently.

From the substation group's perspective, the concern is generally in the maintenance and well-being of assets distributed across a wide geographical plain. Substations can be anywhere and because of this, they are many times unmanned. Meaning that there are many substations which do not have people working in them on a regular basis. This is due in part to the physical location of these assets, as they can reside in locations which may be considered inhospitable for human inhabitant. And these substations will maintain field systems, which are used to manage and maintain assets within the switchyard that the substation is meant to support. From the perspective of this group, maintenance and communications are of extreme importance. If the field systems are not properly maintained and fail, it may take a considerable amount of time to restore or replace them, because of the location of the assets themselves. As a result, maintenance crews and periodic visits to check the current state and conditions of these facilities are required. Communications is of extreme importance to these facilities as well, because it is the only way that they can enable remote support of the field systems assets. Furthermore remote connectivity to these sites is necessary for other business areas within grid operations, so that the necessary monitoring and control capabilities can be used to support overall grid health. The mission of the substation group is then much different from the mission of the control center, generation or metering side of the organization.

The control center group functions in two capacities, one to actually conduct command and control, why the other serves to support the systems that are used to provide this initial function. The team that supports command and control is tasked with actually utilizing the applications and systems needed to support grid operations. They are highly trained engineers and technicians, who understand how the big picture grid functions and how to read the various monitoring tools implemented to provide the current status of grid assets. A fault on a line for example, might be viewed from a control center terminal. The response to such a fault might be for an operator to isolate it and then notify a line crew to physically visit the location and correct the problem. This makes the applications that are used for such monitoring critical and those applications are managed by a separate team that is IT centric, typically referred to as the SCADA group.

The SCADA group is really just an IT organization with specialization in the realm of real-time systems. The specialization extends from the fact that they understand how the applications work and can provide immediate support in the event that one of these critical applications fails. The specialization is important because of the criticality involved in having the operator to be able to respond in near real-time. This means that the IT staff that supports these assets must be able to respond in near real-time if something were to occur. As a result this group represents the first group of people to really respond to a cyber security event and as a result must maintain such expertise. Cyber security is then a very important aspect within the SCADA IT group, as this team of people will be required to identify and eradicate such an event in short order in an effort to ensure the reliability of grid operations.

Power generation is much different from the other groups because the generation facilities are many times owned by multiple utilities or entities, with one being the actual operator. As a result generation facilities are beholden to multiple utilities, increasing the stakes associated with their reliable operation. If something fails it could affect a wide range of customers associated with a wide range of utilities. This effectively means that power plants are really further regulated by the utilities themselves, meaning that failure at one can affect many different utilities, requiring a service level agreement between them to ensure that acceptable measures are taken in the event of an incident. These plants then function like mini utilities themselves and maintain IT groups to support the control system within as well as operators to provide command and control over the plants assets. In addition to this, there are various maintenance groups which are responsible for the plants' hard assets such as the turbine, generator, etc. The control system group, like the operators in a control center, monitor the entire plant to ensure that everything functions within normal operating limits. When something fails, they are tasked with dispatching the appropriate maintenance staff in order to correct the problem.

Smart meter groups are concerned with the health and status of the advanced metering infrastructure. They may be required to manage millions of meters, which represents millions of points, one for each customer. These assets are very distributed because customer locations can exist virtually anywhere. A neighborhood of 200 homes for example may represent 200 meters and a number of collectors implemented to support those meters. When you start thinking about on the scale of a large city like Washington DC, you can begin to visualize the scope of the effort, because every household maintains a meter and the group must be able to support reliable operation of the smart meter as well as be able to acquire information related to billing for each customer. They then must be able to bill and support each customer individually, which is a major task given the identified scope. The smart meter group is then beholden to a set of specific customers who do not really understand how the system works. An outage for example may trigger a number of complaints that may result in action by state and local governments.

The point to all of this is that the missions are different amongst the various verticals within the utility; as such the risk is viewed differently. The control center for example views the loss of an application to monitor assets differently from the manner in which the smart metering group views it. The loss of a single meter for a single customer represents a limited impact to the mission of the smart metering group. But the loss of a system that supports command and control for a plants turbine will have a much larger impact in a power plant. The risk is always relative to the situation, but the key is understanding what the priority of the response will be in each of these verticals. Once this understanding can be established, recovery priorities can be viewed for the utility holistically. This will allow utilities to better respond to a systematic event caused by a cyber security threat. Meaning that there are various impact levels, some which may affect others, a study to understand this would then represent an understanding of the total impact risk to the utility. Each of these verticals is also constrained by different oversight responsibilities. Smart metering groups for example are likely regulated by state and local governments. Whereas the control center that manages the transmission infrastructure is bound by FERC oversight. A generating plant might be regulated by the Environmental Protection Agency (EPA), FERC, as well as state and local governments. Understanding these impacts and the prioritization of them would be key in the development of a grid risk management strategy.

ENDNOTES

1. For more information on FERC's role, see "What FERC does" at http://www.ferc.gov/about/ferc-does.asp.
2. For more information on NERC, see http://www.nerc.com/.
3. For more information on NERC, see http://www.nerc.com/
4. For a discussion of terms used for NERC Reliability Standards, see Glossary of Terms Used in NERC Reliability Standards at http://www.nerc.com/docs/standards/rs/Glossary_of_Terms_2010April20.pdf.
5. For a discussion of terms used for NERC Reliability Standards, see Glossary of Terms Used in NERC Reliability Standards at http://www.nerc.com/docs/standards/rs/Glossary_of_Terms_2010April20.pdf.
6. For a discussion of terms used for NERC Reliability Standards, see Glossary of Terms Used in NERC Reliability Standards at http://www.nerc.com/docs/standards/rs/Glossary_of_Terms_2010April20.pdf.
7. For a discussion of terms used for NERC Reliability Standards, see Glossary of Terms Used in NERC Reliability Standards at http://www.nerc.com/docs/standards/rs/Glossary_of_Terms_2010April20.pdf.
8. For a discussion of terms used for NERC Reliability Standards, see Glossary of Terms Used in NERC Reliability Standards at http://www.nerc.com/docs/standards/rs/Glossary_of_Terms_2010April20.pdf.
9. For a discussion of terms used for NERC Reliability Standards, see Glossary of Terms Used in NERC Reliability Standards at http://www.nerc.com/docs/standards/rs/Glossary_of_Terms_2010April20.pdf.
10. For a discussion of terms used for NERC Reliability Standards, see Glossary of Terms Used in NERC Reliability Standards at http://www.nerc.com/docs/standards/rs/Glossary_of_Terms_2010April20.pdf.
11. For a discussion of terms used for NERC Reliability Standards, see Glossary of Terms Used in NERC Reliability Standards at http://www.nerc.com/docs/standards/rs/Glossary_of_Terms_2010April20.pdf.
12. For a discussion of terms used for NERC Reliability Standards, see Glossary of Terms Used in NERC Reliability Standards at http://www.nerc.com/docs/standards/rs/Glossary_of_Terms_2010April20.pdf.

13. See Rebecca Herold, NISTIR 7628 Guidelines to Smart Grid Cyber Security. Volume II: Privacy Briefing, p. 9 (found at http://www.cpuc.ca.gov/NR/rdonlyres/E222D257-B84F-4575-8DBA-2A55DAFDABA4/0/CA_PUC_day2_Privacy_briefing.pdf).
14. See Rebecca Herold, NISTIR 7628 Guidelines to Smart Grid Cyber Security. Volume II: Privacy Briefing, p. 9 (found at http://www.cpuc.ca.gov/NR/rdonlyres/E222D257-B84F-4575-8DBA-2A55DAFDABA4/0/CA_PUC_day2_Privacy_briefing.pdf).
15. See Rebecca Herold, NISTIR 7628 Guidelines to Smart Grid Cyber Security. Volume II: Privacy Briefing, p. 9 (found at http://www.cpuc.ca.gov/NR/rdonlyres/E222D257-B84F-4575-8DBA-2A55DAFDABA4/0/CA_PUC_day2_Privacy_briefing.pdf).
16. See Rebecca Herold, NISTIR 7628 Guidelines to Smart Grid Cyber Security. Volume II: Privacy Briefing, p. 9 (found at http://www.cpuc.ca.gov/NR/rdonlyres/E222D257-B84F-4575-8DBA-2A55DAFDABA4/0/CA_PUC_day2_Privacy_briefing.pdf).

CHAPTER **3**

Smart Metering
The First Security Challenge

3.1 INTRODUCTION

If there is a front line in Smart Grid, then smart metering would be considered the point where Smart Grid begins. The primary mission of a meter is to monitor power consumption through metrology, which measures the amount of load an entity takes in. Meters are most widely used on commercial and residential buildings. In the past, utilities would physically visit each meter on a schedule in order to read it and determine how much energy a consumer was using. They have come up with complex mechanisms to predict how much energy a consumer would use in order to be able to bill the consumer. One way to do this is to read several meters in a neighborhood and bill consumers based on those reads for a period of 1 year. Companies could then take the average bill and assume that consumers in the same area would basically utilize the same amount of energy. Obviously this process is inefficient as it will and has led to people being billed for energy consumption that does not represent reality. It should be noted that utilities have compensated for this by using complex formulas that are likely considered trade secrets in order to ensure that consumers get a fair shake. Moreover, consumers were likely under-billed in an effort to ensure that the utilities could not be sued for improper accounting practices.

The use of old meter technology has resulted in a loss of revenue for the utilities. If they could somehow gather precise numbers about all consumer use of energy on a grand scale, they would more likely than not realize higher revenues over time. The problem that the utilities have is that they need a smart meter for each customer and even the small utilities have hundreds of thousands of customers. The larger public utilities and independently owned utilities may have millions of customers. Therefore, at the core of the problem are the costs associated with changing out so much equipment (a meter for each customer) and how to build a communications network that supports that many consumers. As a result, utilities have not really deployed smart meters because the cost benefits of the technology may not be realized for years into the future.

3.2 THE COST OF SMART METERING

If the utilities deploy smart meters today, they will be able to gain more accurate meter readings and thus be able to provide more precise bills to their consumers. However, it may take years of planning, strategizing, and implementing such technology. And there is a cost associated with the deployment of each smart meter, which is estimated at anywhere from $50 to $200 per unit. Taking the mean cost estimate, which is around $125, a utility would theoretically have to multiply that cost by the number of meters that it wants to deploy. Moreover, there are other costs associated with building out the architecture, including consulting integration costs, labor related to the installation of equipment, testing to ensure that the architecture functions properly, etc. And then there is the matter of ensuring that the functional requirements of the system have been met. For example, Pacific Gas and Electric (PG&E) has admitted that their smart meter deployment maintained some flaws that resulted in utility customer bills actually increasing.[1] This represents a problem for the utility in terms of costs because of liability issues and general problems with customer service. There is a cost associated with all of this, and that cost represents what will actually be required in order to deploy a successful smart metering program.

Smart meters have existed as both *advanced meter reading* (AMR) and *advanced metering infrastructure* (AMI). AMR meters typically represent one-way communication that allows utilities to gain meter data remotely, eliminating or reducing their need to do manual meter reads. The costs associated with AMR are reflective of the typical costs of the deployment of smart meters. The goal is to get accurate meter data so that precise billing to customers can be accomplished. AMI, however, brings additional functionality to the utility that can even be extended into the actual consumer space. AMI meters typically represent two-way communication that allows the utility to do more than just collect meter readings remotely. And the costs of implementing two-way technologies far outpace the costs associated with implementing one-way communications. The reason behind this lies in the need to provide the capability to allow a central control point to have access to and control every meter within the architecture, while at the same time ensuring that meters have the ability to report information back to the control point. And then there is the need to process this information so that it can be used in billing and other applications that make the meters "smart." These applications are used to implement Smart Grid programs that allow the utility to realize a benefit.

3.3 SMART METERING PROGRAMS

The benefits of smart meters reside in smart meter programs that the utility chooses to implement. The goal of smart metering programs is to change the behavior of the consumer so that the utility can reduce costs and overhead, and improve the quality of service. Most of this stems from the concept of base load. It is cheaper to run a power plant at full capacity than it is to stop it when energy is no longer needed and start it again when power is needed.[2] If the power plant could run continuously and supply the needed power without interruption, that is all power generated by the generation facilities was always sold, then the utility would recognize a significant cost savings in their operation. This of course involves changing the consumer's behavior in favor of requesting power on a basis that can be easily trended by the utility. This trending will eventually allow the utility to predict how much power will be needed on a daily basis. The utility accomplishes this through various programs designed to change consumer behavior, such as

- Demand response (DR)
- Load profiling (LP)
- Automated load control (ALC)
- Customer outage detection
- Customer voltage measurements and power quality
- Distributed generation (DG) monitoring and management
- Remote connect/disconnect
- Real-time meter reading and programming

Demand response (DR) is a system that enables, through economic incentive, the utility to reduce consumption as a means of meeting supply-demand challenges. Instead of monolithic tariffs that mask market prices for electricity from the consumer, DR provides price transparency either through real-time or averaged tiers, allowing the consumer to make informed consumption choices based on real price information. DR programs have shown significant benefits in informing the consumer and driving down peak loads. Without DR, the response to peak load was either a pure supply-side response of construction and generation, or, ultimately, rolling brown-/black-outs during peak usage periods. A successful implementation of DR will allow the utility to control the consumer's use of power, which is far cheaper than changing generation at a power plant. With two-way meter communication, this becomes possible.

Load profiling (LP) allows the utility to measure and collect detailed usage and consumption patterns at user locations. This information is then available for capacity planning and DR consideration. Furthermore, the ability to profile the amount of load needed will allow the utility to determine how much load needs to be generated, thus creating the ability to run power plants at base load continuously.

Automated load control (ALC) provides load management capabilities as a response to the utility's reliability requirements. ALC allows the utility to force load limiting during peak or outage periods to manage load in the face of limited supply. ALC would generally be considered a heavier-hand approach than DR, and is used as a measure of last resort when DR cannot satisfy the load shedding requirements. Combining ALC with LP capabilities allows utilities to automatically control load based on the needs that have been profiled. Computers would be used to determine when the load needs to be controlled as a result of information from LP applications. To put this into perspective, the utility would know how much consumption is necessary on a given day based on the weather, information about each consumer, and trending from previous years. The utility, for example, may believe that it needs to generate at least 2,000 MW on a given day in order to support 500,000 consumers. However, there could be an anomaly that creates a usage pattern change, which is where ALC would come in. Load can be automatically adjusted with a change in consumer behavior. In other words, a computer tells the system where to reduce load and where to increase load based on these changing conditions, usually through DR programs. This illustrates how the three programs work together to make the meters "smart."

Voltage measurements and *quality measurements* provide both the utility and its consumers with key performance indicators regarding power quality. Power quality demands have readily increased as microprocessors and microelectronics have permeated consumer premises over the past few decades. The impact of sags and spikes is greater on modern equipment, and power quality issues can predict wider issues in, for example, equipment failures and performance problems. Power quality monitoring capabilities enable more rapid detection, diagnosis, and resolution of power problems.

The current utility operations model largely depends on consumers to notify utilities of outages due to lack of instrumentation and measurement. The two-way communications medium used by AMI enables real-time evaluation of outages, impact, and ultimately response. An AMI system will participate in a modernized grid to detect and address emerging problems before they impact service. By reading voltage and quality measurements, the utility will be able to detect outages within its service areas before or when they occur. This allows the utility to respond to and manage gyrations within their energy delivery services.

Remote connect/disconnect, also called TFTN (turn-off, turn-on), provides the utility with the ability to enable and disable power at customer locations through remote commands. This enables the utility to manage account changes more quickly and cost effectively, especially in multi-unit dwellings and high-turnover areas, without rolling a truck to the premises to disconnect and connect services. In some instances, remote disconnect can be used to prevent outages by disconnecting circuits where problems have been detected. In addition to this, the utility can remotely enable and disable energy services to the consumer when an order is placed for service.

AMI allows for real-time or near-real-time meter reading and programming. AMI provides time-stamped meter data, allowing utilities to obtain usage information as required for billing, capacity planning, or other processes. Real-time meter reading and programming allows for the utility to provide services remotely to the consumer and aggregate energy usage remotely, which may represent more exact billing information.

The AMI system will allow for measurement of remotely generated power by PHEV, solar, or other *distributed generation* (DG) infrastructure. This system will function by having meters roll backward instead of counting IN and OUT separately. This means that when generation on the consumer's part results in energy being inserted into the power grid, the meter counts the amount of energy released into the power grid versus counting the amount of power that is used. This capability allows consumers to use their excess generation to reduce their own electricity bills.

There are numerous possibilities with regard to programs that can be implemented as a result of smart meters. Various scenarios exist that can be applied to this concept, and smart meters can be used to accomplish a never-ending stream of goals. The implementation of smart meters then represents business value associated with the Smart Grid. Figure 3.1 illustrates one possibility as to how smart metering technology can be implemented for some of the programs just described in an interconnect fashion.

The utility company controls and aggregates information from meters. This allows the utility to profile load usage based on consumption levels reported by each meter. The goal of this would be to align consumption with the base load output from their generating facilities. If consumption stays consistent with the base load, then no action is needed and the utility would simply monitor power quality to make sure that nothing unusual occurs. If power quality ever indicated a disturbance, then the utility could initiate its outage management processes to predict and respond to failure before it begins. One control that a utility would be able to use in such an event would be automatic load control. This would be true if the system detected the need for more load

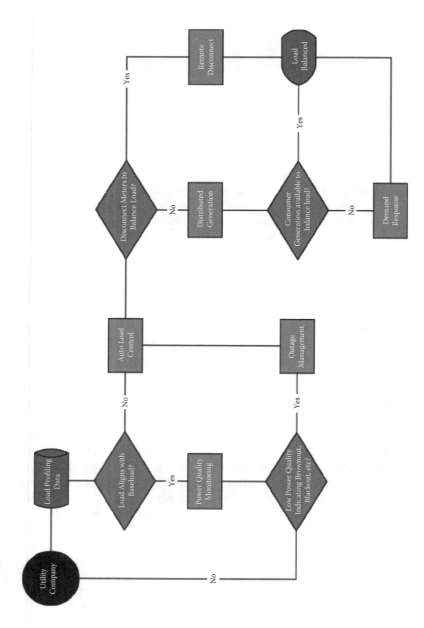

FIGURE 3.1   Smart metering program flowchart. (From SAIC - Josh Wepman.)

based on the detection of low power quality. Auto load control would also be used under conditions where the utility wants to automatically ensure that consumers are only using the energy from base load output. To adjust power, the auto load control feature might disconnect some circuits or use access into the home area network to reduce power consumption. If DG were available, the load could be adjusted by taking in energy from the consumers themselves, if they were signed up to introduce power into the grid. The point of this exercise is to ensure that there is a balanced load and that the utility only has to run base load power.

3.3.1 The Smart Meter Architecture

AMI is an architecture that provides two-way communications for conducting advanced metering functions. The system is based on the standard metering communications architecture, built using the ANSI C12.18, C12.19, C12.21, and C12.22 standards. The standards provide a communications framework for initiating and receiving information in a two-way metering communications environment. The AMI system is designed using the following components:

- Meter
- Collector
- Meter control system
- Meter data management system

These components rely on additional network architectural components such as

- Wide area network (WAN) communications
- In-home display
- Utility demilitarized zone (DMZ)

Figure 3.2 represents a graphical depiction of the AMI architecture.

The architecture begins at the customer premise and ends at the meter data management system (MDMS). Everything inbetween the customer premise and the MDMS represents the metering architecture. This is because the goal in AMI is to provide the consumer with energy from the distribution system, while controlling the use of energy through the AMI architecture. Some of the components within the architecture can be developed by a single or multiple vendors. Generally, meter vendors will provide the meter, collector, and control software, while others will provide the WAN communications, in-home devices and communications, as well as the MDMS. All these technologies are integrated in a

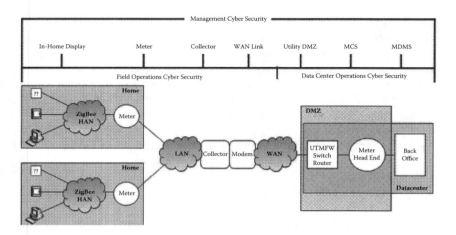

FIGURE 3.2   AMI architecture. (From Josh Wepman, SAIC. PowerPoint presentation, unpublished.)

manner that supports the control and monitoring of energy consumption, while managing the distribution of power to the consumer. The following subsection describes each technology type in greater detail.

3.3.2 In-Home Display

The in-home display provides an interface to the customer premises. The customer premise consists of equipment that is connected to the in-home display. The in-home display can measure the consumption of the customer premise equipment and will one day provide consumers with load control over their equipment. The in-home display is the component used to provide customers with information about their own energy consumption and will eventually provide consumers with IP connectivity to their home in an effort to control energy consumption. This means that the consumers will be able to theoretically control their pool pump from a central console within the home, or log in to that device over the Internet to turn lights on or off, adjust their AC settings, or simply monitor their energy usage remotely.

The in-home display maintains two interfaces: one to the meter and the other to the customer's premise equipment. The interface to the meter is used to transmit information about customer energy usage to the meter and back to the utility through the AMI network. Currently, it does not provide the utility with control capabilities over customer premise equipment; however, it is intended to provide the utility with the ability to conduct demand response. This interface can use various wireless frequencies, but it is generally accepted that the meter vendors will determine which frequency will be used to connect the in-home display to the meter. For example, the in-home display might use a 900-MHz radio frequency so that it can send information to the meter and the meter can send signals to it. It is this communications medium that will provide the utility with the ability to talk to the in-home display.

The second interface on the in-home display is used by customers to control their premise equipment. ZigBee, which is an 802.15.4 wireless communications medium, is quickly becoming the wireless standard for in-home communications from the display to consumer assets. 802.15.4-2006 specifies the physical layer and media access control for low-rate wireless personal area networks (LR-WPANs).[3] The standard is intended to offer a type of wireless personal area network (WPAN). This is done by focusing on low-cost, low-speed ubiquitous communication between devices. The emphasis of the standard is on very low cost communication of nearby devices with little to no underlying infrastructure, intending to further exploit this to lower power consumption.

Consumer assets represent appliances within the home that consume energy, where management of that energy can result in a cost savings to the consumer. For example, the air conditioner and furnace are used to cool and heat a home. With a ZigBee interface, consumers will not only be able to control the temperature from a console, but will also be able to remotely manage their furnace and air conditioner (assuming they have the ZigBee interface) remotely. Or consumers might provide control of their air conditioner to the utility so that it can be shut off in the event that load is needed in another area of the distribution system. This control by the utility represents demand response as a program that can be used to manage energy consumption.

Another key feature that the in-home display brings courtesy of the ZigBee protocol is for consumers to actually see how much energy they are consuming in real-time as a dollar amount. This enables features such as prepayment of electricity, or for the consumers to adjust energy consumption behavior because they recognize that they are experiencing a high cost during any period of time during the month.

3.3.3 Smart Meters

Smart meters are used to collect information in regard to energy usage, as well as to conduct control services such as circuit disconnection. While energy usage is a measure of consumption, circuit disconnect refers to the meter's ability to break an electrical circuit between a building and the distribution system. Meters are smart because they provide the utility with the ability to interact with them from a centralized remote location, via two-way communications, and they provide the ability to enable IT systems to make programmed automated decisions on their behalf. This means that the meter can effectively shut off electricity at a site, if it is commanded to do so, either via control by a human or automatically as a result of an event by a computer.

Smart meters use the ANSI C12.18 protocol for sending and receiving instructions. The protocol standardizes the manner in which it passes information between meters and collectors. In addition to passing information using this protocol, a 900-megahertz radio frequency (RF) mesh network is used to actually move payloads from one device to another. The radio frequency provides the transportation medium in which the meter can pass its "table data" (i.e., C12.19 packaged data). Both smart meters and collectors each have interfaces that are dedicated for communications among themselves. The C12.19 table data is what passes over the RF between the meter and the gatekeeper. This standard is used for the purposes of ensuring commonality between what is sent and received by the smart meters and collectors. For example, if a meter wanted to send an acknowledgment to another meter, it would organize the "ack" into a table. The table would contain the "ack" in a specific column, and the protocol would basically be telling the receiver to read this line first. This represents the ANSI C12.19 protocol, which is the "packaged data." Next, the smart meter would send the table to its communications interface, which would get transmitted across the mesh network. Sending the table to its communications interface would be like the meter saying that this package is ready for transmission, illustrating the C12.18 protocol process. Then the data would actually be submitted through the mesh network and when it reached the next meter, it would understand what it was receiving and tell itself to open the table and read the "ack" line. This may be a simple approach to message transmission, but it is an important one because the simplicity of the message exchange becomes extremely important in order to scale millions of meters.

3.3.4 Neighborhood Area Network

The neighborhood area network (NAN) is also referred to as the 900-MHz mesh network.[4] This network is probably the most critical component in smart metering because it is what connects the communications network associated with the smart meters. That is, the NAN is what makes it viable to connect smart meters via two-way communications. This is because it is cheap and very scalable. Whereas doing the whole thing over Internet Protocol (IP) communications would become extremely costly given the number of households. This is not to say that a NAN could not be done via an IP network, because as the costs associated with it decrease, it will likely become a viable reality. IP-based meter-to-meter communications will likely be regarded as next-generation Smart Grid technology.

The 900-MHz mesh network works similar to Cisco's Open Shortest Path First (OSPF) protocol. Basically, a meter sends information using the radio frequency for layer 2 for transportation of the information to the next meter. The next meter may not be the intended destination of the information and may in fact be a stop along the way. All meters work as relays (like little routers) that relay the information received to their closest neighbor, until the data reach a collector that can talk C12.21 or C12.22. This works, in that the meters send the data using the C12.19 protocol, which is primarily for meter-to-meter communications. But it may say that the destination of the package sent is to a meter that can talk C12.22 so that the information can then be passed across an IP network. The point of the mesh network, however, is that the relaying of information between meters is how one would reach a certain meter or get information from another meter. As long as the meters are able to communicate with one another via the 900-MHz RF, then two-way communications to and from the meter are available.

Security applied to this radio frequency includes what is commonly referred to as *spread spectrum*. Spread spectrum works by taking the signal generated by the smart meter into a particular bandwidth, and spreading in the frequency domain, resulting in a signal with a wider bandwidth. Frequency hopping is a method of transmitting radio signals by rapidly switching a carrier through many frequency channels, using a random sequence known to both smart meters involved. Spread spectrum then provides the meter communication signal sent with the ability to use a wider bandwidth, which means that each packet is sent across a much larger range of radio communication signal. An example would be to break up the table package and instead of sending it over one line, it would be sent over 1,000 channels to the next meter, where it would be reassembled by the frequency. This is done to make it more difficult to intercept the package in transit. Frequency hopping comes into play by randomly switching the channels sequentially in a manner that only the smart meters communicating with it know. For example, instead of sending the package across the same 1,000 channels, it will randomly select different channel paths that the next meter also knows. This effectively means that security is being performed by hiding the way that the data from point A get to point B.

3.3.5 Smart Meter Collectors

As stated previously, smart meters eventually connect to smart meter collectors. Smart meter collectors are generally the same thing as smart meters; however, they also maintain the ability to communicate across a wide area network, usually using the IP. This is where the world of metrology gets transitioned to the world of information technology (IT). When packages finally reach the collector through the mesh network, they have arrived because the meters intended to send them on to an IT device such as a meter control system or meter data management system. The important thing to note about the collector is that there is usually one for every predetermined number of meters. And the collector can be a meter among meters as well as perform the collecting function. It has an interface that communicates with the mesh network and another interface that communicates across the wide area network (WAN).

The C12.21 protocol is used where collectors utilize telephone modem-based communications to get data across a WAN. A telephone modem may be built into the collector or it may be connected to a dial-up modem, depending on the preference of the utility. While the C12.22 protocol is used where collectors utilize data communications networks, a broadband card or leased line may be used to transmit the C12.22 protocol information. In either case, the point of the protocol is so that the information can be transmitted across the WAN communications link that is available. For example, you would not be able to transmit a C12.18 package across a C12.22 data network if C12.21 has been specified for transmission of the package. The protocols simply tell the collector how to handle the information and how it should be transmitted to the next destination; this is usually managed by an information technology group within the utility.

3.3.6 Wide Area Network (WAN)

As the C12.18 protocol addresses optical port communications, the C12.22 protocol addresses Internet Protocol- (IP-) based communications. C12.22 is primarily used as the protocol for WAN communications within the AMI architecture. The C12.22 protocol sets the standards that the collector utilizes to communicate with the meter control system. The meter control system usually resides in a demilitarized zone (DMZ) on the utilities data network, and data collected and aggregated by the collectors is passed through this WAN network to the meter control system for additional processing. The WAN usually transmits data through a third-party telecommunications provider using whatever the utility decides for a medium, whether it be Broadband, Frame Relay, CDMA (code division multiple access), GSM (global system for mobile communications), etc. This would be the equivalent of attaching a wireless air card or DSL (digital subscriber line) modem to a collector, addressing it, and routing the traffic to the DMZ, where the meter control system receives and processes the data.

Traffic from the collector is routed through the telecommunications provider's network until it eventually reaches an *access point*. The access point within the telecommunications provider's network may then have an established virtual private networking (VPN) tunnel to a firewall within the utility's DMZ. It is done this way to ensure that the IP addresses from the collectors are not routable outside the telecommunications provider or utility data networks. However, this means that the security of the data and the commands issued as a result of the meter control system depend on how good the telecommunications provider is.

The use of technologies such as CDMA further restricts what a utility can do to manage communications security. However, this technology provides the collector with the greatest amount of bandwidth available through the use of 4G communications. One of the primary concerns from a security perspective with this technology is whether or not the communications frequency can be compromised or intercepted by an external threat. CDMA is a wireless technology that has, to this date, never been compromised, meaning that the mechanism used to transport communications from the modem to the telecommunications provider is considered fairly secure. However, GSM, CDMA's counterpart, has been hacked. Furthermore, the attack involves spoofing a cell tower that results in a man-in-the-middle attack on the traffic between the modem and the telecommunications provider. This demonstrates that it is only a matter of time before the threat sources catch up with the technology, which means that the utilities will have to ensure that they utilize the latest and greatest available—that may be a costly proposition considering the number of collectors that it would take to support an entire utilities smart meter deployment.

Leased lines and or dedicated fiber owned by the utility are another option that can be used to bring communications from the collector to the utility's central offices. The use of utility fiber probably represents the greatest amount of bandwidth available and the most secure means of communications. However, the cost may far outweigh the benefits in this case, leaving the option of leased lines or DSL from the collector to the utility. In this case, traffic would still travel to the telecommunications provider from the collector, but the utility would gain assurance with regard to security because wireless is not used. Moreover, this option lends itself to a long-term investment that may not require modifications whenever technology is breached by threats. The utility would not have to replace modems because the wireless technology has been compromised, for example.

3.3.7 Utility Demilitarized Zone

The *demilitarized zone* (DMZ) is a public-facing area within the utility that is used to handle interactions with the public data networks, or the Internet. The DMZ is needed because the information from the collectors is technically coming from a data network outside the utility. For example, if information is coming from a collector to the meter control system over a leased line, then the information is coming from the telecommunications provider and not the utility itself. This third-party communications method is necessary because the telecommunications provider is generally regarded as the entity that can provide WAN support at the most reasonable cost. The Internet Assigned Numbers Authority (IANA) has assigned port 1153 for TCP and UDP data communications using the ANSI C12.22 protocol. This is important because utilities will have to open port 1153 TCP and/or UDP in order to receive messages from their collectors. An example firewall security policy within a utility would be to "allow" traffic from a pool of collectors with IP addresses, which are public, through the firewall using the ANSI C12.22 protocol over port 1153 to the destination of the meter control system.

3.3.8 Head End System

The meter control system is used to manage the AMI architecture from a central source. The meter control system is the system that can initiate commands to all components within the AMI. The meter control system is housed in the DMZ, which means that it faces the public Internet. A firewall is used to isolate IP-based communications to and from the meter control system. Furthermore, the firewall usually provides a VPN link to the telecommunications provider. The data from the collector are passed through this VPN in order to get it to the meter control system. The meter control system uses this VPN connection to initiate commands to the AMI as well as to receive data from it.

3.4 SMART METER AUTHENTICATION

The most common authentication solution in the WAN relies on token-based authentication. The collectors and meters maintain a security token that is used to verify their authenticity. Before a command can be issued on the meter or gatekeeper, from the meter control system, the token is verified by the authentication system. If the token is valid, authorization to initiate a command on the collector or meter is authorized by the meter control system.

Device-to-device authentication is an important aspect in the AMI architecture because it is the primary mechanism used to ensure the integrity of the architecture. Older versions of smart meters utilize weak encryption algorithms and can potentially have an impact on availability due to the authentication architecture's scalability. The more handling of authentication required by the smart meter architecture, the more likely it is that it will fail. This is why weaker encryption was initially used in smart metering architecture, in order to sustain reliability. As time has gone on, vendors have realized the need to enhance security through the use of stronger encryption algorithms. The stronger the encryption algorithm, the less likely a threat source might be at injecting malicious code into the meter. This is because to be successful, the threat source will most likely have to defeat the authentication mechanism, which is much more difficult as the algorithm gets more sophisticated. While this reduces the risk that field components, which are physically available to virtually any threat source, will be manipulated by or injected with malicious code, it also potentially creates reliability issues within the architecture. Because more processing power is needed, which means that meters have to be more sophisticated, the more expensive it will become to deploy smart meters and there is the potential that they will be less reliable at the same time.

The fundamental issue in this architecture is that AMI relies on the availability of the meter control system, and the security of the individual meters/collectors depends on the protection of the authentication medium (i.e., the tokens). Furthermore, as more meters and collectors are deployed, the meter control system will be impacted with additional overhead that may eventually cause reliability problems.

Another solution being promoted and tested is the use of SEED-based key exchange.[5] The SEED-based key exchange allows meters to simply understand the algorithm needed to decrypt a key without having to logically store the key in its memory. This means that key-based authentication can be used without requiring that keys be stored on meters and collectors. The meter control system serves as the key master, as it maintains all the keys logically and all components are required to authenticate to it prior to granting access or allowing for session initiation. The system would only allow one-way communications down to the meters and collectors. This means that the meters and collectors could never initiate a request unless the meter control system told them that they had to do something; that is, they would never act independently. This scheme works by allowing a user to log in directly to a meter control system using basic operating system authentication, which provides access to the meter control system application.

The LAN (local area network) key is derived by the meter-unique meter identifier and the "Meter SEED." For authentication, the SEED value is sent to a device, and the device creates its unique key derived from the SEED and the unique device identifier using a proprietary algorithm. This allows for each meter key to be unique. The Meter SEED is a value only known by the meter and the meter control system.

For example, if a user at the meter control system wishes to access a meter, he or she would place a request for access (i.e., attempt to log on). The meter would request an authentication code, which can be used in unison with the meter's SEED file to decode the key of the meter control system. In doing this, the meter is able to authenticate the user or device prior to allowing a function to occur. The security portion of this lies in the key-based authentication. This solution provides near-foolproof non-repudiation of key values and thus a secure way of authenticating device-to-device communications. Furthermore, it provides resilience in that the SEED can only be modified if the password on the meter control system is modified. This means that every time the master password on the meter control system is changed, the SEED is changed on all meters, in an effort to update them to the fact that there is a new authorization process in place. This is done without ever having to change the key values.

The collectors talk C12.22 protocol to the meter control system (C12.21, if an analog modem is used), but cannot initiate a connection directly to it. The meter control system polls the collectors, makes a request, and data are submitted back to the collector using the C12.22 protocol. Because the protocol requires encryption, all the table data are encrypted back to the meter control system. The SEED value is used as the primary key management mechanism and can be changed at the meter control system, which affects every SEED value in the architecture. When the password on the meter control system is changed, the SEED is automatically updated on all collectors and meters. The collector also maintains the algorithm to build a key in real-time anytime the meter control system requests authentication.

The meter control system builds the key based on an algorithm that only the SEED can decode. Once the key is built, it is sent to the collector, which uses its SEED to decrypt the key and it acknowledges itself with a message back to the meter control system, which essentially gives the meter control system the green light to send a request. This function is similar to that of a state table in a stateful firewall. Once the collector makes that acknowledgment, the session can begin, and it begins with the meter control system saying "send me your latest table." The collector responds with an acknowledgment, and it is processed as long as proper authentication has occurred. This happens for every single transaction.

3.5 SMART METERING SECURITY

It is worth noting that AMI was developed so that communications and the functionality of meter data acquisition and control could be improved. As communication and functionality improve for the utility, so does their profit margin. This is primarily because efficiencies are realized through a centralized command-and-control interface for AMI. Cyber security, however, plays an important role in this space because cyber security risks threaten the utility's ability to realize a return on the AMI investment.

Utilities and meter vendors have typically tried to protect the security of AMI by hiding what they do and how they do it. As metering does not traditionally involve IT, it is not generally thought of as a technology to exploit. Furthermore, it is thought of as something that resides on a closed network and is not visible on the Internet. Most utilities will tell you that the vendors are responsible for the security of their AMIs, and most vendors will tell you that they have built all the proper security into their technology to mitigate the cyber security risks. While all this is true, it is being explained in general terms, which glosses over the reality, which is that the protection of this infrastructure is the job of everyone involved. And it is the utility's responsibility to lead this effort and manage how their vendors secure their platforms, how those platforms will be managed securely, and to ensure that all cyber security risks identified have been mitigated. It is important to understand that throughout this chapter we have been discussing different technologies produced by different vendors that can be organized in many ways to produce the same result. This means that a single vendor cannot ensure security because it depends on how the utility implements its product and with which other vendors its product will communicate.

3.6 SMART METER VENDOR MANAGEMENT

Because vendors are on the front line of security, it is important that they are appropriately managed by the utility. Utilities will then want to ensure that they procure some cyber security expertise in order to facilitate procurements with their vendors. This expertise is usually found in-house in an IT department that is responsible for enterprise cyber security issues. The compliance component of this team is what should be sought after because they can analyze the security of vendors to determine if they are in alignment with best practices. However, as smart metering technology is new, there is not a whole lot of regulation around the security associated with it. In addition, the components—meters, collectors, etc.—are not computers, so standard IT security frameworks can generally not be applied. The question then becomes: How do you secure an infrastructure for which there are no security standards?

Utilities could bring in penetration testers to test their implementations, which would more likely than not uncover cyber security issues with the deployment. The problem is that penetration testing is not an exact science; it all depends on what the ethical hacker is looking to accomplish on any given day. This really means that the external penetration test will uncover current cyber security risks, but maybe not the future concerns. Penetration testing should then be used only if there is a real concern about the way something is supposed to be secured in regards to the vendor's product. Otherwise you can bet that the vendor will procure expert penetration testers to determine how good their product's security is. The results will likely be used to improve security over time. A utility can take a principled view of this by requiring that the vendor have a third party to conduct a penetration test of each component procured and that the results of such are released to the utility. This would integrate parts of the security lifecycle with the utilities project lifecycle, baking security into the project.

There is still the problem of not having any regulations or requirements that could be applied to the architecture in order to gain some assurance as to the cyber security posture. National Institute of Standards and Technology (NIST) has been developing a document entitled NISTR 7628, which looks at the interoperability and cyber security of the overall Smart Grid, taking AMI into account as an important component of the new grid. And there is the SG Security Working Group sponsored by the UCA International Users Group that meets quarterly and has various projects that utilities can participate in to understand smart metering. This group has even produced a set of standards based on the Department of Homeland Security's "Catalog of Control Systems Security: Recommendations for Standards Developers"[6] and NIST SP 800-53. Still, the requirements attempt to apply traditional IT security concepts to technology that does not fully use IT. The concept, however, is sound if you look at the principles represented in the requirements and look at how they are applied in smart metering.

Access control is a security principle that is applied to security in general. Banks store money in vaults and provide a large door with a combination to control access to it. This is in comparison to putting a firewall in place to block all traffic, except for what is authorized to a specific destination. This same concept can be applied to smart meters. Why can't a utility ask its meter vendor how they control access to each meter within the mesh network? Another question that might be asked is how does the vendor identify and authenticate meters communicating with meters? Or if a meter does get hacked, how will the utility know what happened? Will logs be available and, if so, where are they stored? These are important questions to ask a meter vendor because it forces them to reveal their security posture. If the vendor tells the utility that they do not control access to meters because they cannot, doesn't that mean that anything on the mesh network can talk to anything? And that raises the following question: What stops one from accessing the mesh network?

Vendors should be asked security questions by utilities to understand the security posture of what is being procured. The problem is that utilities do not know what to ask because there is no standard surrounding smart meter security. However, smart meters are being deployed aggressively across the country, so something should be done. Remember that these are major capitalized investments that must be in place for at least 10 to 20 years, so utilities should want to know whether or not they need to replace all their meters if a worm develops for a meter and the meter cannot be upgraded, so that the threat can only be mitigated through replacement of the meter to a newer model with a feature that allows for remote upgrade of firmware.

Going back to NIST SP 800-53, there is a list of eighteen control families, all of which could be used to ask meter vendors security questions about their product. As an example, Table 3.1 illustrates how this questionnaire might be framed and what a response from the vendor might look like.

TABLE 3.1   Smart Meter Vendor Questionnaire

Security Control	Description	Vendor Response
Access Control	How is access controlled from one smart meter to another?	Smart meters can only be accessed through the mesh network.
Identification and Authentication	How do users uniquely identify and authenticate themselves to smart meters?	Each meter has a unique identifier.
Audit Logging	What security events are logged by smart meters?	Smart meters notify the meter control system when they are not available.
Communications Protection	What protects communications from one smart meter to another?	Spread Spectrum frequency hopping is used to protect meter to meter communications.
Configuration Management	What are the configuration options within the smart meter and what have you implemented?	All the standard options are available in the meter.

The utility's enterprise cyber security team may be able to evaluate the vendor's responses and develop another set of questions to further clarify the security issues as they perceive them. For example, if smart meters can only be accessed from the mesh network, what is the access control mechanism that prevents access to the mesh network? Here the utility might assume that the risk is reduced because the mesh network might be secured, but the vendor has not actually determined this; they have simply passed the risk onto another function. Managing vendors in this manner will allow the utility to reduce risk through question and answering. Moreover, for the answers that cannot be determined, the vendor can order additional testing to verify whether or not the risk really exists.

3.7 SMART METER SECURITY MANAGEMENT

Security management is a principle where the utility should embrace the idea that it needs to at least try to prevent things from happening and have the ability to respond if something does occur. Vendor management is an example of attempting to prevent something from happening. A smart meter incident response practice might be an example of having the ability to respond if something does happen. The security of products and services is only a piece of the security puzzle. The vendor can make all the security features in the world available to the utility, but if they are not used or managed properly, then they may be useless. For example, a vendor may include in its product the ability to securely access the meter control system application via a username and password. But the utility might decide, for the sake of efficiency, that it does not want its employees to take the time to enter a username and password. The control was available, but the utility failed to use it. This is an important concept to understand given the scope of smart metering and the fact that there are millions of points of presence in this architecture.

Because of this vast scope, it will be important to manage security via a structured uniform approach. The utility should then determine through risk analyses what is needed and then how those needs can be applied. Going back to the vendor management questionnaire, which controls are available and unavailable should be determined first. A process for managing the implementation and operation of the controls that can be implemented should be established. Meters, for example, might generate and report logs that could be used to determine whether or not a security event has occurred. At some point, meters may be neither controllable nor report information back to the meter control system. This could be because the mesh network has failed or it could be because a threat has pulled the meter out of its socket. If the mesh network failed, it should be investigated to establish the root cause to determine if it failed because of operational limitations or because communication was interrupted by a threat source. If the meter is pulled out of its socket, then it is possible that threat has done this so that they can disrupt power or to study the device to see what could be done to the meter. Just as an IT department would want to know whether or not a laptop was stolen, the utility should want to know if a meter was stolen. These are all examples of security events that might occur and should be managed according to security principles.

NIST SP 800-53, SG Security, and a number of other standards, requirements, etc., maintain the concept of operational management. This concept is used to manage the actual operation of the metering infrastructure. For example, if something happens, an incident response practice should be used to investigate. If a disruption of service occurs, there should be a disaster recovery plan to recover from the incident that caused the disruption. These operational management controls should be mandated by higher-level policy so that the organization understands what must be done. The reality is that the metering department within a utility can probably leverage the enterprise cyber security department to help in the development of procedures designed to manage security with regard to operations.

As the smart meter network is complex, and uses various protocols that are usually not known by general IT practitioners, operational security management of the metering infrastructure may be a new domain that requires a combination of metrology experts and cyber security experts to determine how best to manage security. Factoring in what the products deployed can and cannot do should result in the utility's ability to manage most of the risk associated with its smart metering deployments.

While policies are simply paper-based statements that require actions, they do not directly address security concerns. They basically should state what the utility must do in order to mitigate potential risks. And because no formal requirements exist that are generally accepted to mitigate the risks associated with AMI, it will be important that policies be developed cross-functionally and with input to what is available in order to implement mandates that protect a utility's AMI deployment. One such policy might be that the utility requires that vendors be evaluated for cyber security during the procurement process, or that the solution deployed use a standard and available centralized identity management solution. To measure operational security effectiveness, the utility will have to understand what the risks are, and those risks will need to be analyzed to determine how they should or could be mitigated.

3.7.1 AMI Vulnerabilities

Advanced metering infrastructure (AMI) maintains three inherent vulnerabilities: (1) components are physically available to anyone, (2) physically available components communicate to IT systems through authorized access points, and (3) utilities are reliant on telecommunications providers for protection from collectors to the meter control system(s). These are considered inherent vulnerabilities because they are the most obvious; and without countermeasures to mitigate the risks identified in each, they are easily exploited.

Smart meters, NANs, and collectors are all physically available to anyone in the general public. Anyone can walk up to a meter or collector within any given neighborhood and attempt to do something to it. The concern is not so much of the physical threat of ripping a meter off a pole and taking it out of service, although this is something about which the utility should be concerned. The concern is more of what could a cyber-threat do to a meter if it were able to plug a laptop into it. Because smart meters are programmable, viruses, malware, etc. can probably be written for them. And with meters being physically exposed to anyone, what is to prevent a cyber-threat from engaging a meter and installing malware directly? Looking at IT systems on corporate networks, one of the threat vectors is through a USB device with malicious code on it. As smart meters maintain optical ports that could be accessed with a laptop, the question becomes what controls would prevent a threat from doing this?

Authentication to the meter would prevent a threat source from engaging a meter physically with a laptop and installing something on the meter. If the meter required authentication to directly engage it and the threat source was unable to authenticate, then the threat source would be unable to access the meter. This is one layer of protection that prevents access. However, more protection is needed, seeing that the meter can be repeatedly accessed by the same threat, which could eventually brute-force the meter into allowing it access to the meter. Potentially, notification of invalid authentication attempts to an alert console would provide additional protection so that this physical threat source could not just repeatedly attempt to gain access to the meter. In reality, the application of technical controls should be applied to the meter to determine what a threat source might attempt to do and then remediate the risk through the implementation of controls.

Probably the most likely attack scenario that utilities can expect is that threat sources may attempt to jam the mesh network radio signals, making the mesh network fail. While this is not too much of a concern today considering all the smart metering programs are really not critical to ALC and other smart applications, it is still something to think about. Failure of the mesh network already appears the most significant problem facing smart meters today. That is because there are a variety of things that can interfere with the radio signals, causing signal degradation and even failure of the entire mesh. These problems generally can be corrected on a case-by-case basis using root cause analysis. If the cause of the network failure is due to someone jamming the signal, then the utility might want to consider putting in a communications monitoring system that looks for rogue signals and alerts whenever any sort of jamming is detected. The detection of such an event does nothing to stop it, but this information can be given to the vendor, who might be able to help identify solutions to correct such problems. The other problem with the mesh network concerns whether or not the signal can be breached and the network hacked.

If the mesh network is hacked, a hacker may be able to then impersonate meters, injecting all kinds of code and malware into the network. Furthermore, this could cause integrity problems that may result in the utility not being able to trust the information provided by the smart meters. Imagine if hackers started selling discounts on electricity, similar to the way that some have defeated satellite TV's defense to give away free cable. It may be possible for someone to buy a 75 percent discount in electric service simply by having a hacker change his information related to usage, for a small fee. Another possibility might be for a hacker to hack into an smart metering infrastructure and send a blanket disconnect command, telling all meters in the service area to disconnect power, creating a small localized blackout. This would be a small problem that the utility could easily solve, but what is the impact of an event like this on the utility's reputation and ability to keep the lights on? The point is that the mesh network must be protected from unauthorized access and the spread spectrum frequency hopping technique may not be enough. The need for some sort of encryption of traffic may be necessary in order to prevent frequency intrusion. The problem is that the mesh network is already fragile, so adding something like encryption may cause a larger problem. Nevertheless, utilities should look at what they can do to further secure their radio transmissions between meters.

Finally, if meters are hacked, a problem might stem from the fact that they have access through the utility firewall on TCP and UDP Port 1153. While there are currently no known hacks for this port and protocol, as smart metering grows in popularity, it is likely that utilities will face attacks directed at this kind of traffic. If a utility, for instance, ran communications from collectors directly to firewalls using the standard method of communications with no VPN, it may be possible that the collectors could be directly engaged by threats on the Internet. The point is that once the collector is accessed by the threat, there will be open access through the firewall to the meter control system. The meter control system represents the largest risk because if it is compromised, then it can be used to compromise the entire AMI infrastructure.

3.7.2 AMI Impacts

The impact to AMI represents an impact to the business function for which it is being implemented. Utilities see the following programs as direct reasons for implementing AMI:

- Demand response
- Load profiling
- Automated load control
- Customer outage detection
- Customer voltage measurements and power quality
- Distributed generation monitoring and management
- Remote connect/disconnect
- Real-time meter reading and programming

If these programs do not function properly, then the utility will not get a return on investment from smart metering and thus will lose money. As a result, utilities will want to protect their investment by ensuring that these impacts do not occur. Currently, utilities do not see how these impacts could be realized because so far, nothing has happened. But as time goes on and smart metering gains more importance and visibility, we will begin to see attacks on it.

If the utility cannot take advantage of its demand response program because a threat has compromised its data, giving inaccurate reads on what the demand is versus the supply, the utility will likely end up making a decision that may affect millions of people. If the utility reads that no more load is necessary, when in fact a great deal of load is needed, what will happen? Generally, bad things happen at utilities when decisions are made based on inaccurate data. Electrical engineering is almost an exact science; if the load provided is not enough to compensate for demand, then consumers will likely experience electrical problems.

If the profile of the load used by personnel is inaccurate, then the utility may forecast or make strategic decisions that will cost the utility millions of dollars in lost revenue. Consider that load profiling at the meter level determines how much energy their generation facilities need to produce in order to keep everything in balance. Meter information can have a direct impact on generation, and the costs associated with ramping generation up and down are costs that the utility cannot afford to take lightly. Moreover, inaccurate load profiles may drive decisions about time of use, etc., which could impact other customers.

One of the scariest of all the scenarios is in the automated load control program. This is because it has direct ties to the distribution system, where decisions will be made by smart applications when load is to be automatically controlled based on events that occur. What will happen if a threat source changes conditions such that the distribution switches load to compensate for something that did not really occur? And what if meters are told to disconnect by a worm or virus? Among all the services AMI offers, the disconnect function is the most controversial in information security circles as it is the only one that directly controls the flow of power to the home or business. While DR and ALC involve sending a signal to a meter that could result in switching off an appliance, the consumer is usually able to easily override such action. However, absent some rewiring, there is no equivalent override for the disconnect switch. In fact, one of the purposes of the disconnect switch is to ensure that customers who do not pay their bills are denied electricity until they do. They are typically targeting areas that are highly transient, such as college communities or in low-income areas. The greatest concern is that a successful attack could allow someone to gain control of the head-end system and switch off power to thousands of customers all at once. In addition to causing widespread blackouts, repeatedly switching the power off and on could create frequency imbalances and surges in the grid that could damage loads and destabilize the entire grid, potentially causing damage to generators, transformers, and other equipment in the path. Such a consequence would be much more severe than a simple power outage, resulting in damage to expensive equipment with replacement times of more than a year in some cases. Effectively taking temporary control of the meter network could lead to widespread power outages lasting weeks or perhaps longer. That is something we have not experienced in North America since electricity became ubiquitous 70 to 80 years ago. However, the biggest problem in all of this is typically the reputational harm that the utility will experience as a result of a cyber-attack. And it is better for the utility to attempt to mitigate those risks now, making a fair investment in cyber security, rather than to wait for something to happen. When an attack occurs, it is not going to be the scope of the attack that harms anyone or any utility; rather, it is going to be the fact that it happened, the uneasiness that will be felt by the consumer, and the mandates that will surely be legislated by Congress.

When the Internet started, there really were no viruses. They were being written and they were infecting machines, but there was no real impact. It was not until people realized that their identities were being stolen, as a result of these viruses, that anti-virus became a must. The same thing goes for the patching of systems; previously it was something that was done to enhance the consumer's experience. Once worms started taking down e-mail servers and businesses' services, patches became extremely important and now businesses are more vigilant than ever in this regard. This was, as discussed in previous chapters, the information security age, where the impact was in data loss. Today, we are still fighting that battle, and at the same time a new battlefield is emerging.

Cyber security as related to the utility field is currently a place where "information can now be used to control physics," as Joe Weiss of Applied Control Solutions puts it. The manipulation of data can be used to turn off electricity or to steal energy. There will be multiple impacts that can be realized as a result of cyber security risks and smart metering. But the paradigm change is that now the hackers can actually do something to harm human life. As we get into other chapters, readers will soon realize that we may indeed be entering the world of "skynet," as process control systems are nothing more than robots being controlled by computers. And to make the process control systems more efficient, we are going to have to make them more intelligent. This is done by automating more and more processes. The reality is that there is risk that must be mitigated in smart metering and, to accomplish this, the concept of cyber security must be applied to the program life cycle in place for smart meters.

ENDNOTES

1. Martin LaMonica, PG&E admits to flaws in some smart meters. CNET news, May 11, 2010, retrieved from http://news.cnet.com/8301-11128_3-20004645-54.html.
2. Cristina L. Archer and Mark Z. Jacobson, Supplying Baseload Power and Reducing Transmission Requirements, retrieved from http://www.stanford.edu/group/efmh/winds/aj07_jamc.pdf.
3. For more information on the protocol, see LAN/MAN Standards Committee of the IEEE Computer Society, Part 15.4: Wireless Medium Access Control (MAC) and Physical Layer (PHY) Specifications for Low-Rate Wireless Personal Area Networks (WPANs), IEEE-SA Standards Board, June 2006, retrieved from http://standards.ieee.org/getieee802/download/802.15.4-2006.pdf.
4. Other methods exist in AMI deployments to transport data from meters. They include power line carrier, cellular, and some existing radio networks. However, 900-megahertz-based wireless communication is by far the preferred method in North America. Power line carrier communicates over the same power line that the electricity flows but typically requires bypass wires at the transformer. Additionally, power line carrier is known to induce radio frequency emissions that disrupt other wireless communications. Some wireless carriers are marketing cellular-based meter communication similar to how mobile devices communicate over the Internet. These deployments are limited as of the publication of this book. Because

they rely on the availability of cellular towers, some rural areas may experience coverage problems. At the moment, utilities using this approach are required to pay monthly fees akin to what cellular customers pay for their phone service. While the rates per meter are a fraction of what cellular phone users pay, the aggregate cost can be significant. Finally, Sensus, an AMI vendor, uses a network called FlexNet that operates on a different frequency that typically has a wider range per base station. More information can be found at http://www.sensus.com/flexnet/SUS-1004_FlexNet_Brochure_seperated.pdf. Solutions based on WiMax and other technologies are also being considered. All these options have the typical trade-off between cost and performance, with security levels varying based on the technology and individual implementations. Because the 900-megahertz option is the most common, this book is focusing on that one.

5. SEED is an encryption algorithm developed by the Korea Information Security Agency in 1998. More information can be found at http://rfc-ref.org/RFC-TEXTS/4269/kw-seed.html.

6. See http://www.us-cert.gov/control_systems/pdf/Catalog%20of%20Control%20Systems%20Security%20-%20Recommendations%20for%20Standards%20Developers%20June-2010.pdf.

CHAPTER 4

Home Area Networking
Giving Consumers Control or Opening a Pandora's Box?

4.1 INTRODUCTION

As briefly discussed in Chapter 3, the home area network (HAN) is where the Smart Grid connects with the consumer. It is the part inside the home or place of business, and it is the part over which a utility or other service provider has the least control. This is hardly a surprise given that businesses have been operating a less high-tech version of the HAN for quite some time. In larger facilities, these are typically called building management systems (BMSs). These systems offer a variety of measurement and control functions for heating, ventilation, and air conditioning (HVAC); water; lighting; and other environmental functions. Typically, building security systems are operated separately, but theoretically they are just another type of building automation. Increasingly, these systems also are monitoring and controlling the same types of end devices that utilities manage. For example, manufacturing facilities are full of programmable logic controllers (PLCs) that may operate valves, control the frequency of motors, and coordinate complicated assembly line functions. Because of the amount of energy required, an energy management system, which is one of the roles a BMS plays, is typically involved to measure performance and track efficiency. This can offer companies an early warning for equipment that is about to break down or incur higher costs just like lower fuel economy in one's car may be an early indicator that an oil change is needed or that the tires need more air.

For the home, building automation is relatively new. With the possible exception of a programmable thermostat, most consumers do little to automate the monitoring and control of their energy usage. Hobbyists have played around with technology from companies such as X10 for more than a decade, and security system vendors have offered cool-looking tools to turn on and off lights and arm their homes directly

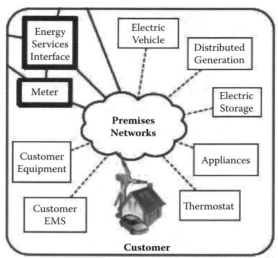

NIST Smart Grid Framework 1.0 January 2010

FIGURE 4.1 The home area network (HAN). (From NIST Framework and Roadmap for Smart Grid Interoperability Standards, Release 1.0, p. 35. Retrieved from http://co llabora te.nist.gov/twiki- sggrid/pub/Smar tGrid/IKBFramewo rk/NISTFr amework AndRoadmapForSma rtGrid-Interopera bility_Relea se1final.pdf.)

from their mobile phones. However, none of these efforts has seriously tackled energy efficiency across wide geographies, largely because electric utilities have mostly chosen to not to engage consumers. To most utilities, their involvement stopped at the meter. How a consumer used their electricity was not the utility's problem. And while the notion of a smarter grid was not inspired by a desire to implement some "nanny state" philosophy on how we use electricity, it did acknowledge that consumers need help. Specifically, they need information to make and automate decisions about their electricity, including the price and how much electricity an appliance or task will consume. Knowing that electricity is $0.10 per kilowatt-hour is of little use if a consumer does not know how many kilowatt-hours a load of laundry requires.

In essence, this is what the HAN is designed to provide. However, after that, things get a little more complicated, as Figure 4.1 demonstrates. The HAN, after all, sits at the crossroads of several different markets. We first have the electric utility market that is concerned with enabling a utility to send a pricing signal to a HAN device and to possibly control its operation to some degree. We also have the consumer variant of that market that includes companies such as Google, which will receive the same pricing signals along with appliance data in order to help consumers make better decisions through the data analytics tools they offer. There are also the appliance manufacturers that will be embedding communications chips into their products in order to facilitate the monitoring of electricity usage by a device in the home, over the Internet, or over the AMI (advanced metering infrastructure) network. Similarly, the nascent and frequently overestimated home automation market is hoping that this will be their chance to convert every home into something straight out of a *Jetsons* episode that would include integration of devices and data from the utilities, appliance manufacturers, data analytics vendors, and various other service providers to improve energy efficiency and effectively become a home energy management system. Finally, we have a cast of players still yet to come onto the scene. They include those pitching distributed energy resources such as solar panels, wind turbines, batteries for storage, and even plug-in electric vehicles that will be discussed in more detail in later chapters. However, for purposes of the HAN, they represent devices that need to be controlled and monitored by the homeowner, the utility, or by a designated service provider. Because decisions on the devices to be used and the method of monitoring and control will be made by consumers, utilities, or other service providers, it is likely that the architecture of HANs, their operation, and the components they contain will vary dramatically even within the same neighborhood. This means that everyone will need to consider security issues very carefully and use caution when relying on security controls implemented by another party.

4.2 ELEMENTS OF THE HOME AREA NETWORK

As just discussed, the actual architecture of the HAN and its components is likely to vary considerably. Nonetheless, a variety of groups have sought to define various elements of the evolving Smart Grid and solidify some use cases. The OpenHAN Task Force, which was created by the SG Systems Working Group under the Open Smart Grid (OpenSG) Technical Committee of the UCA International Users Group, has developed the *UCAIug Home Area Network System Requirements Specification*[1] that identifies twelve different device types.

As depicted in Figure 4.2, they include

1. Energy services interface (ESI)
2. Utility ESI
3. Programmable communicating thermostat (PCT)
4. In-home display (IHD)
5. Energy management system (EMS)
6. Load control

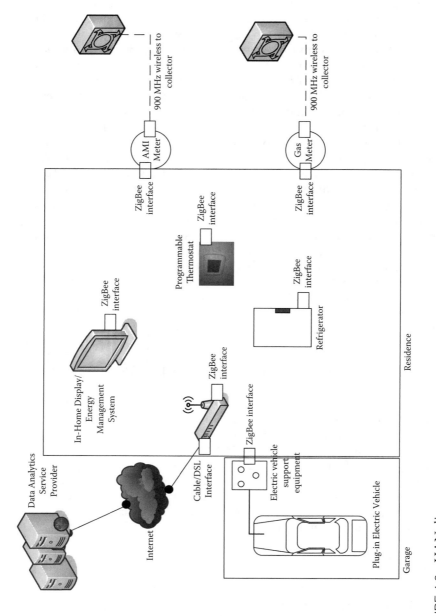

FIGURE 4.2　HAN diagram.

7. AMI meter
8. HAN non-electric meter
9. Smart appliance
10. Electric vehicle supply equipment (EVSE)
11. Plug-in electric vehicle (PEV)
12. End-use measure device (EUMD)

While not all these devices will exist in every home or business, many will be present and others, such as distributed energy resources (e.g., windmills and solar panels), may join the list in the future. What becomes clear is that as Smart Grid technology takes shape, the HAN will quickly become a very sophisticated network and, consequently, cyber security will become increasingly important.

Let's now consider each element of the HAN individually.

4.2.1 Energy Services Interface

The ESI is perhaps the first part of the HAN deployed as it is often represented by a chip on the AMI meter. In many implementations, this is a ZigBee[2] chip, which we discuss in more detail shortly. The ESI interface is effectively the gateway that routes data between two networks. With a utility that implements AMI, the demarcation point is easy—the meter. With other providers that deliver their services over the Internet, the demarcation point may be more difficult to determine or may not exist at all. In the Utility ESI example, it is critical that the types of traffic flowing from the AMI network to the HAN and back be closely controlled. If the utility has too much access to the HAN from the AMI network, then a consumer's privacy is potentially violated. If there's too much access in the other direction, then a malicious customer could attack the meter network or something further upstream. That is why a clear process must be implemented to identify which devices in the HAN the utility can communicate with and register them accordingly. In other cases, the communication may be a simple broadcast of price changes and the utility would not need to treat the device any differently than a radio station treats a radio. It should ultimately be up to the consumer which devices he or she chooses to communicate with the utility and which may leverage alternate service providers. In the case of outbound communications, the utility should be clear about what kinds of communications it receives and what it trusts. At a minimum, the HAN chip on the meter should only be allowed to upload data onto the meter that another process on the AMI network retrieves. Additionally, utilities must validate the data provided and never assume that it can be completely trusted. Instead, it should establish tolerances for the data

ranges expected and discard or manually verify any data that falls out
of that range.

4.2.2 Programmable Communicating Thermostat (PCT)

For many consumers, their first direct interaction with the Smart Grid
will be through their thermostat. While sophisticated in-home displays
(IHDs) and iPhone-enabled monitoring and control features have seen
some limited deployment, consumers can already buy programmable
communicating thermostats (PCTs) that can communicate with Google's
PowerMeter online service that provides consumers with data on their
energy usage. The thermostat is also the most logical place to begin. It
typically controls heating and air conditioning; and when they are both
generated from electricity, they are generally the biggest users of electricity, giving consumers the biggest opportunity for savings on time-of-use
rates or other incentives if the thermostats are programmed correctly
and respond to price signals. For utilities, the value of the thermostat
will be to make slight alterations in consumer behavior to—hopefully—
collectively avoid the significantly higher costs at peak times where more
costly generation sources, such as diesel, might be required. Unlike peak
shifting, where a consumer is encouraged through price incentives to
delay washing dishes or laundry until the evening, adjusting the thermostat would basically ask people to forego a little comfort by raising
the temperature when air conditioning would kick in and lowering the
temperature for heat in the winter. To consumers, a few degrees is hardly
noticed, but to a utility, it's just enough load reduction to forestall the
use of more costly generation or to have to build new generation plants.
While utilities often benefit financially from a new coal or nuclear plant,
they only do so if the demand for that energy is consistent throughout
the day and throughout the year. Smart Grid technology helps to make
that happen.

The security risks, at first glance, would appear minor as one is
simply sending a signal to a thermostat that normally can be manually
overridden. However, it is important that a utility get reliable results
when a demand response event occurs. If, for example, the thermostat
were to report back to the utility that it was adjusting the temperature
level when in fact it did not, then the utility may rely on that report
and other similar reports from other residences to plan for less generation. At the moment, such false information would likely not have much
immediate effect as the utility typically makes generation requests based
on aggregate energy use measured at the substation level and further
upstream. Moreover, the meter would still measure the correct energy
usage. However, as utilities begin to rely more on their alternate sources

for long-term planning and possibly short-term projections, unreliable responses from thermostats could have some impact. Similarly, if an attacker could target thermostats from the AMI network, the Internet, or wirelessly through ZigBee, then false price signals and other malicious commands could be sent to the thermostats. Without effective authentication mechanisms, one could adjust the thermostat to raise or lower temperatures to an unacceptable level and might even lock out manual changes. This could have severe health consequences to the young, old, and infirm. Consequently, it is important that mechanical overrides be built in and that consumers understand the risk that these remotely controllable devices present. Those risks are manageable and are currently very low. However, people have a tendency to over-rely on technology and put more trust in it than they should, so let this be a warning to stay vigilant.

4.2.3 In-Home Display (IHD) and Energy Management System (EMS)

While theoretically the in-home display (IHD) and energy management system (EMS) could be considered separate devices, it is likely they would be integrated together. The goal of the IHD is to extend the services that are already available through PCTs on the market and extend them to allow a single device to monitor all "connected" devices in the home. In theory, a ZigBee chip could be installed on a wireless network switch to allow access via a personal computer or even a smart phone with the appropriate software installed. However, the IHD is being marketed as a more streamlined appliance that might also offer weather updates, cooking recipes, and Web access via a touch-screen panel installed in the kitchen or family room. It could display projections of energy usage based on weather forecasts, past behavior, and current temperature set points, and could recommend changes in response to time-of-use rates, demand response price signals, or possible inefficiencies discovered, such as frequent defrost cycles for the freezer or above-average number of on and off cycles for heating and air conditioning that may reflect poor insulation or doors and windows being left open. The security of this device would clearly be critical from the consumer's perspective in both preventing others from snooping as well as preventing someone from using that device to manipulate household appliances. Attacks could be launched wirelessly through the ZigBee or similar short-range protocol, through the AMI network, or over the Internet. Additionally, because utilities and third parties will likely bundle Internet access as a potential marketing hook, the device will also be exposed to potential malware when a customer is surfing the Internet, which also means that there must be a mechanism for frequent security patches.

4.2.4 Load Control and Smart Appliance

While most of today's Smart Grid focus has been on AMI and some lip service paid to the HAN via PCTs, the HAN won't be much of a network until multiple devices operate on it. This will only happen if load controllers or smart appliances with load controlling functionality are deployed. Like other HAN devices, load controllers and smart appliances would have wireless communication chips, such as ZigBee, built into them to communicate with the ZigBee chip on the AMI meter, IHD, or EMS. However, appliance vendors have been understandably reluctant to deploy communications chips in their products until there is both a demand and agreement on the communication standard to be used. For example, a representative from Whirlpool commented at a conference in 2009 that such chips would not be incorporated in their product line until at least 2015. Moreover, the chip is only one part of the issue. There is still a question of how much information consumers will want to know about the relative cost of using a device under a particular rate scheme. Simply knowing that the rate is $0.25 per kilowatt-hour would not be that helpful. Instead, they may want to see a display on their washer that tells them that a load of laundry will cost $5.00 now but $0.50 in 3 hours. Perhaps the IHD could provide similar data but that requires familiarity with a wide array of appliances and of course consumers would have to remember to check the IHD before doing the laundry.

The load control category is really a subcategory of a smart appliance and, in some ways, a temporary phenomenon until we actually have smart appliances. Today's load controllers may be nothing more than a box that connects between the appliance and the electric outlet and can switch the power off and on when triggered remotely. Not only is this inappropriate for sensitive appliances that require a more graceful power-down operation, but they also fail to take advantage of energy-savings features that could be leveraged in smart appliances. For example, a smart freezer could, on request, alter its freeze-and-defrost cycle to correspond with time-of-use rates.

For both smart appliances and load controllers, the security challenges are similar. People have been hacking into radio-controlled toys for decades, and it is likely that hobbyists will have their fun with smart appliances. That is why utilities should not rely too much on data they receive from these devices and it is why appliance makers should implement mechanical protections to ensure that someone cannot issue a wireless signal to disable safety shut-off features that prevent appliances from overheating and starting a fire. While it is unlikely that life support system manufacturers will make devices that participate in Smart Grid energy efficiency programs, many may have wireless communications capabilities for other reasons and should similarly exercise caution. Energy efficiency, or peak shaving, should not be a reason to ignore safety. Additionally, consumers should always have the ability to manually override any automated energy-savings feature. Additionally, consumers should never completely trust the IHD that may provide diagnostic information on their appliance. There is no replacement for occasional visual inspection of critical features of the device.

4.2.5 HAN Nonelectric Meter

While by definition the Smart Grid means electricity, AMI technology is really a communications technology that could be applied to other metering functions, including water and gas. Due to the different physical properties of water and gas, there is less need for features such as demand response. While shortages do occur, they are not as time sensitive as electricity. For example, a water shortage may develop over weeks and months and not minutes or hours as is likely with electricity. Time-of-use billing does not help much. However, there are still efficiencies to be gained by performing remote meter reads, particularly if there is an infrastructure already built out to support electric meter reading. Additionally, giving consumers real-time access to usage information can be very useful in helping to control costs and waste. For example, if daily usage is available, a consumer may be able to discover a small gas or water leak. Integrating a nonelectric meter with the HAN can present challenges, particularly if the meter is controlled by a different company from the electric meter. Each utility will want to verify that the other has performed appropriate due diligence with respect to security and that the IHD or EMS is designed to support nonelectric uses. Because gas-powered furnaces in colder climates account for such a large portion of a consumer's energy budget during the winter, it seems inevitable that gas usage will be monitored someplace within the HAN.

The security risks posed by nonelectric meters are not much different from those of electric meters. They can be manipulated to commit energy theft and may be capable of turning off the flow of gas or water remotely. Consequently, the guidance provided in Chapter 3 applies here as well. Integration with the HAN could present some risks, but for now it appears manageable, particularly given the fact that it will simply monitor usage with fewer control functions beyond the thermostat.

4.2.6 Plug-In Electric Vehicle (PEV) and Electric Vehicle Supply Equipment (EVSE)

In Chapter 10 we discuss plug-in electric vehicles (PEVs) in much more depth, as their widespread adoption will fundamentally change how we think about electricity and the Smart Grid. Simply put, the power alone required for millions of PEVs would require significant changes to our electrical grid. Consequently, it is important to understand how this technology will interface with the HAN. While the business models and infrastructures are still being developed, they generally fall into two camps. The first would treat PEVs like any other load on the grid. The person who owns or leases the property associated with the electrical outlet pays the bill and can independently submeter a vehicle charging station and bill per charge just like the cell phone charging stations that are popping up at airports and hotels. For residential users who do not have a submetering capability to measure the cost of electricity for the charge, they will just have to guess if a visiting friend or relative offers to pay the cost of charging one's vehicle. The other camp would treat the PEV like a roaming cell phone and allow it to plug into any outlet with the vehicle owner getting billed for the electricity used. The latter option would be difficult to implement based on the current state of the grid and probably is not sufficiently beneficial for utilities to implement. We'll touch on these two business models more in Chapter 10, but the significance for the HAN is really what, if any, new services it needs to provide for PEVs. Under the first scenario, it is just another load. Under the second scenario, it may need to relay identifying information about the PEV over the HAN and AMI network in order to bill the right account. The second scenario is certainly fraught with fraud issues as someone can simply try to impersonate someone else's PEV and associate all its electricity costs to the phantom PEV.

Most likely, the HAN communication associated with the PEV will come from the electric vehicle supply equipment (EVSE), which will likely be responsible for controlling the flow of electricity to the PEV and possibly from it in a scenario where the PEV acts as a distributed energy resource (DER). Either way, the EVSE would be responsible for determining the appropriate window for charging. Because PEVs draw a large amount of electricity, any EVSE will need to consider both time-of-use rates as well as the residence's overall power at any given time. Under some scenarios, the electric company may deliver a separate circuit just for vehicle charging but in the meantime, a wide variety of interim solutions will be needed. Under the most likely approach, the EVSE will act just like a load controller with some added features for detecting the PEV and providing possible slow or fast charging options. The EVSE may also report through the HAN back up to the utility that vehicle charging is ongoing. Under this scenario, there is little additional risk posed by the PEV. However, if the EVSE miscalculates the power required and the other current and planned electricity demands during the charging, the home could blow circuit breakers and potentially cause damage to sensitive electrical equipment. Consequently, the EVSE communications must be secured, most likely at an assurance level higher than that of other appliances.

4.2.7 Mobile HAN Devices

Consumers are already taking advantage of applications that can be used with their smart phones. These devices will likely be used as a mobile interface that can control the various components within the home area network. Consumers may even decide that it would be beneficial to use these devices to adjust settings in their homes remotely. When the HAN becomes a reality, we will likely see a number of applications for it. This appears to be one area of concern when addressing cyber security implications of using a HAN. The specific question that remains to be answered: How will mobile devices interface with the HAN in a manner that provides adequate protection?

The number of applications available for smart phones today is staggering. Moreover, these applications can be developed by virtually anyone and loaded onto any smart phone. Some vendors, such as Apple, perform some vetting of the application's security before it is made available in the online store, but other vendors are not doing any security reviews. While commercial products may not always have the best security, having an established company that risks reputational damage, or worse, is at least some incentive to implement some due diligence with respect to security. For many mobile applications, the barrier to entry into this market is so low that virtually anyone can develop and sell these applications. And while such democratization of software development can be viewed positively in terms of innovation, it does have some disadvantages with respect to accountability for what is written. Rather than resist this movement, it is important that HAN device vendors recognize the threats presented from these unknown applications just as Internet-facing applications have had to resist attacks of all kinds. Just as utilities must be somewhat suspicious of data coming from the HAN and therefore have restrictions on how it is used, HAN devices must also limit what mobile applications can do. For example, a consumer should be able switch lights on or off and adjust the thermostat within a certain range. However, someone should not be able to exceed certain thresholds remotely that could lead to pipes freezing or pets dying of heat exhaustion. Moreover, a remote application should not be able to update firmware or use the HAN to communicate with the utility's AMI network or further upstream. Additionally, one could envision some sort of certification program for mobile applications accessing the HAN where a HAN device not only authenticates the user, but also effectively authenticates the application by confirming that the application software certificate is registered with the HAN. This could help prevent malware from gaining a foothold in hundreds or thousands of HANs that still could pose a problem even if the HAN limits functionality of mobile applications.

Beyond energy monitoring and control, the HAN presents numerous opportunities for home automation that are very attractive to consumers, and remote access via smart phone applications will certainly be part of it. However, the same thresholds and common-sense thought processes should be applied. For example, does someone really need to open his or her garage door or unlock the front door from hundreds of miles away? While one can envision some applications, it might just be simpler to give the neighbor the key to check on a pet or water the plants and limit the remote part to turning off the alarm system. The goal with any automation process is to automate the good without automating the bad. That means turning lights on or off remotely while not alerting thieves to the fact that one is not home or offering them an easy way in. There are potentially hundreds or even thousands of risks associated with this. As the technology becomes more advanced, utilities and consumers will likely want to pay attention to what they use for controlling components connected to their HAN. Mobile device security will then play a very important role in ensuring that millions of consumers are protected from various threats as their use increases the risks associated with them.

4.2.8 Other Devices

The HAN could easily morph to support a wide range of other devices. This could include DERs such as wind turbines and solar panels, which are discussed in more detail in Chapter 7. While distributed energy at home is far from being a widespread phenomenon, when they are deployed, the HAN may likely play a key role in relaying generation capacity to the electric utility in order to balance all generation resources at all levels of the grid. This will be a tremendous challenge to both standardize the communication across a wide array of both sophisticated commercial providers and relatively inexperienced residential providers who are simply offering excess capacity on an occasional basis. Because data coming from the HAN may not always be trustworthy, utilities will need to approach any data about generation capacity with caution—not only because the data could be manipulated by individuals seeking to either harm or defraud the utility, but also from individuals who are not sufficiently competent in operating a generation resource that may be relied upon by the larger community.

4.3 HAN COMMUNICATIONS

Because the HAN is largely a wireless network of devices that are both provided by a utility and purchased by consumers, it is extremely important that the industry adopt a communications protocol that can support these devices, can support new technology as it evolves, and can be deployed securely. As noted above, ZigBee is the most commonly proposed communications protocol for the HAN. It is based on the IEEE 802.15.4 standard with additional layers built on top of it. As of this writing, it is currently under revision. However, the legacy version 1.03 is currently the one being deployed, and it has a number of security challenges among other issues. To keep the protocol lightweight, so that chips with limited processing power and low cost can be used, the specification takes some shortcuts with security. In his writings and talks, Joshua Wright of InGuardians describes the many vulnerabilities found in a typical ZigBee device and ways to exploit them.[4] He describes how over-the-network provisioning of encryption keys is done initially in plaintext, that there is very weak protection against replay attacks, and that encryption keys are shared among devices in a HAN, so if one device is breached, they all can be breached.[5] This is why security-conscious organizations like the U.S. Military have resisted suggestions to implement the ZigBee for their Smart Grid deployments, noting that in its current implementation, it is not sufficiently secure for controlling loads. However, that hasn't stopped ZigBee from being deployed in manufacturing plants, home automation systems, and environmental control systems for commercial buildings. Like all technologies, sound risk management methodologies must be used to determine whether the risk is worth taking. For a building management system, that risk may be worth it as a building engineer can monitor for anomalies and replacing it with new technology within a given building is not difficult. However, relying on homeowners to detect anomalies or replacing devices deployed in millions of homes when the risk becomes too great is normally not a sound business or security practice. Other protocols and specifications have their own challenges. Bluetooth, for example, has its own security challenges, uses too much energy, and would be problematic with larger networks. WiFi, or 802.11 wireless, is certainly ubiquitous and has adequate security for this purpose, but it is currently too costly to be deployed on the kinds of devices being proposed for ZigBee. One option for many homes, which is being promoted by the HomePlug Powerline Alliance, uses the existing electrical cabling in the home to provide a communication path for the HAN. This is very similar to power-line carrier technology discussed in Chapter 3. For a variety of reasons, including the frequent noise that appliances create over power lines, the HomePlug standard is not as popular even though network speeds can be as fast as 200 megabits per second and some degree of physical security in the location of the electric cables provides some additional protection from eavesdropping and manipulation of the data. Despite the high bandwidth and available infrastructure, the medium is susceptible to line noise, making it less reliable without some degree of compensation.[6] Moreover, the wiring inside structures can vary with age and design, and may be less desirable for broad-based deployment.

As noted earlier, communication between HAN devices is only part of the story. The Internet could end up being the primary means of communicating with the HAN as consumers look to household names such as Google to provide energy analytics based on the input they provide. Ultimately, people may end up trusting Google more than their electric utility company. Regardless, the Internet path may be less secure than the AMI network, which will need to account for alternate paths into the home. Attackers should not be able to break into a consumer's home and then work their way up through the AMI network. While currently that would be highly improbable, it is still important that we stay vigilant and not succumb to the pressures to control costs by skimping on security. For example, some vendors have proposed to allow HAN devices like the in-home display to talk directly over the 900-MHz AMI network without using the home meter as a gateway. If that were done with a display that communicated at both 900 MHz over the AMI network and 802.11 wireless home network through the home Internet connection, we could be looking at a potentially dangerous scenario. While most utilities would probably not provide in-home displays that support both frequencies, there would undoubtedly be pressure from consumers and electronics manufacturers to support such models that would also likely provide other desirable features, such as remote monitoring via a smart phone. It would seem that the best advice is not to put oneself in a position where one is likely to be vulnerable to peer or consumer pressure. Consequently, the goal should be not only to pick technologies that are secure, but also to choose solutions whose natural evolution, based on public opinion and cost, continues to favor security.

4.4 HAN COMMISSIONING, REGISTRATION, AND ENROLLMENT

From a security perspective, one of the biggest challenges is figuring out how these vast arrays of HAN devices, some owned by the utility and some by the consumer, can possibly become part of the same network. The OpenHAN Task Force breaks down this process into three parts. The first is *commissioning*. This "allows devices to exchange a limited amount of information (e.g., network keys, device type, device ID, initial path, etc.) and to receive public information. This process is initiated by the Installer powering on the device and following the manufacturer's instructions."[7] Next is the *registration process*. It involves "authentication and authorizing a Commissioned HAN Device to exchange secure information with a specific ESI and other HAN Devices Registered with that ESI. Registration creates a trust relationship between the HAN Device and that ESI and governs the rights granted to the HAN Device in that ESI's network."[8] Keep in mind that after this process, the HAN device is only communicating with other HAN devices in the home or business and possibly with the electric meter that would typically relay information from the utility. In many cases, without enrollment, the utility is not aware of any HAN device and, for privacy reasons, would normally be precluded from receiving information on the existence or performance of any HAN device. However, data from these devices might still be fed into a third-party analytics engine to assist customers with their energy usage. That service might even be offered by the utility, but it is for the benefit of the consumer and is not tied to any utility program. Moreover, the data going to this analytics engine will likely traverse the Internet rather than the AMI network.

The third part of this process is the most significant from a security perspective. It is the *enrollment process*. "This process is only applicable when the Consumer wants to sign their HAN Device up for a specific Service Provider program (e.g., demand response, PEV special rate, prepay, etc.). Enrollment is required for the Service Provider to get specific device addresses or device information. In this process the Consumer selects a Service Provider program and grants the Service Provider certain rights to communicate with or control their HAN Device."[9] In a typical scenario, a consumer enrolls in a utility's demand response program where the consumer agrees to limit or reduce his or her electricity usage during times of high demand or lower generation availability. This may mean agreeing to a higher temperature before one's air conditioning kicks in during a particular time window. The utility would send a signal to the HAN device, such as the thermostat, and the HAN device would acknowledge that it has responded in accordance with what the consumer had previously agreed to do. While the terms of the program may vary, consumers would typically be given a discount on their electric bill for participating, with the amount of the discount possibly dictated by the amount of energy reduction they are willing to accept. That typically translates into agreeing to a higher temperature in the summer and a lower one in the winter, depending on the energy source used for heat. For the utility, the security challenge may be to ensure that the consumer sends back information that accurately reflects the temperature state of the home or the operational state of the HAN devices. Because the consumer may be the one supplying the device, the utility has a limited ability to ensure the device's reliability. At best, it can certify particular devices to be used and rely on the protections that the manufacturer has put in place with the assumption that only a small percentage of the customers will seek to alter them. Currently, many pilot programs call for utilities to supply the devices to maintain and ensure reliability and interoperability. However, as these devices proliferate, pressure from manufacturers and consumers will likely force utilities to accept devices that they have not supplied, particularly if manufacturers are willing to submit to some sort of certification program.

Another security risk concerns manipulation of the device by utilities. This could involve malicious individuals seeking to switch off power to thousands of homes at once, a scenario already discussed in Chapter 3 in relation to the disconnect switch in many smart meters. However, a less sinister motivation might come from the utility wanting to generate more savings from its demand response program. For example, a utility might conclude that it needs the threshold when air conditioning kicks in to be a higher temperature in order to meet the demand on a particularly hot day, and so despite a consumer agreeing to making that temperature 81 degrees when demand is high, the utility chooses to override that setting and make it 82 degrees. In some cases, there may be a tremendous incentive for utilities to make such minor adjustments because the aggregate effect could be large. The utility may even rationalize that consumers would not even notice the difference, and in many cases, they would be right. However, such rationalizations are not only illegal, but they could be dangerous to people with sensitive health issues. Moreover, rationalizations have a way of getting out of control where one degree becomes two and eventually becomes five or six. It also can severely damage trust and cause individuals to withdraw from such programs en masse, a consequence that could lead to blackouts and other disruptions due to a lack of energy. That is why the settings on HAN devices should only be controlled by consumers with read-only access by utilities to verify compliance. Delegating such access to a third party, particularly one that has interests that are not completely in line with the consumer, is, at best, problematic.

4.5 DEFENSE-IN-DEPTH AND OTHER SECURITY SOLUTIONS

HAN security is clearly an evolving concept, in part because use cases and architectures are still very much in flux. Without understanding the assets involved and the potential impact if they are compromised or mis-used, it is difficult for security professionals to fully define the protections needed. Additionally, any HAN security solution must take into account the scale. Every home and business would potentially use the solution, and so cost is a major concern. Placing an intrusion detection system in every home would hardly be economical or effective. Additionally, any device that is designed to look for suspicious events must report its find-ings somewhere. No utility has the resources to handle the volume of traffic that would result, particularly all the false positives. Automation may make that possible. However, it is likely that humans would still be required to step in. Additionally, HANs will likely be subject to technol-ogy convergence where video, audio, Internet, environmental, and other home automation domains come together. Utilities may have to play a part in a much larger market and collaborate with manufacturers, cable companies, phone companies, data analytics providers, and other third-party service providers in order to define the right balance of security and convenience.

Recently we have seen interest in digital provenance at the chip level. That effort would seek to impose additional assurance that the devices and the chips they contain are authentic and not subjected to manipula-tion. Additionally, technologies implemented at the chip level can be implemented to strictly limit the activities that can be performed on the chip, a concept sometimes referred to as application whitelisting. That means that HAN devices could offer a level of security independent of how the consumer chooses to deploy them with attempts at manipula-tion effectively invalidating the device. Many other decisions will depend on trust models implicit in the individual use cases. In general, those trust models should be aligned with the interests of the parties involved. Where someone has a compelling incentive to cheat, adding more secu-rity controls is less effective. This is no truer than with the HAN where, almost by definition, utilities have much less control. As noted earlier, utilities can sponsor certification programs that steer consumers to buy-ing products that have been subjected to extensive security testing and review. Additionally, utilities need to promote best practices in HAN implementation, particularly if consumers are expected to do it them-selves. Over time, we may see HAN firewalls and intrusion prevention products. Ultimately, however, the HAN is part of a much larger uni-verse of home automation solutions that we are only starting to envision. Principles of defense-in-depth and least privilege will undoubtedly be important. Consumers cannot afford to give service providers (including utilities) too much control, and service providers must limit the amount of trust they accord to any data they receive from a consumer's HAN.

ENDNOTES

1. UCAIug Home Area Network System Requirements Specification. OpenHAN Task Force formed by the SG Systems Working Group under the Open Smart Grid (OpenSG) Technical Committee of the UCA® International Users Group, version 2.0, August 30, 2010, retrieved from http://osgug.ucaiug.org/sgsystems/openhan/Shared% 20Documents/OpenHAN%202.0/UCAIug%20HAN%20SRS%20 -%20v2.0.pdf.

2. "The term "ZigBee" and various designations, logos, icons, and graphics are registered trademarks of the ZigBee Alliance and can only be used in accordance with ZigBee Alliance guidelines." See http://www.zigbee.org/imwp/idms/popups/pop_download.asp? ContentID=6700 at p. 9.

3. For a full description of the ZigBee protocol, see ZigBee Smart Energy Profile Specification, ZigBee Alliance, December 1, 2008. It is also worth noting that ZigBee is one of many protocols designed to operate over IEEE 802.15.4, such as the IPv6 low power over wireless personal area networks (6lowpan) protocol. The Internet Engineering Task Force's (IETF) Request for Comments (RFC) 4944 discusses the protocols in detail at http://tools.ietf.org/html/rfc4944.

4. See, for example, Joshua Wright, KillerBee: Practical ZigBee Exploitation Framework or "Wireless Hacking and the Kinetic World," retrieved from http://www.willhackforsushi.com/presentations/ toorcon11-wright.pdf.

5. This also raises the issue of key management, which is a particularly challenging issue with respect to HANs. Consumers want a solution that is maintenance-free and does not require constant care and feeding for things like key revocation and renewal. We will return to key management later in this book as we sum up the security challenges for the Smart Grid going forward. The Smart Grid presents possibly the largest public key infrastructure (PKI) management challenge we have ever seen. Added to that challenge is the fact that there will be multiple PKI deployments at various levels of the grid that may be controlled by different parties that will need to work together. The HAN may end up being its own enclave for many functions, but it will need to talk with entities outside the home that require some form of public/private key cryptography to be used. To get security right, effective key management will be a big part of the ultimate solution.

6. For detailed description of the technology and some of its challenges, see Srinivas Katar, et al. Harnessing the Potential of Powerline Communications using the HomePlug AV Standard. RF Design, August 2006, pp. 16–26, retrieved from http://rfdesign.com/mag/608RFDF1.pdf.
7. UCAIug Home Area Network System Requirements Specification, pp. 32–33.
8. UCAIug Home Area Network System Requirements Specification, p. 33.
9. UCAIug Home Area Network System Requirements Specification, p. 33.

CHAPTER 5

Distribution Automation
Moving from Legacy to Secure

5.1 INTRODUCTION

For many in the electricity industry, distribution is the "holy grail." Unlike transmission and generation, distribution is typically not redundant and is the source of part of the grid where outages occur, such as from downed tree limbs and other weather-related events. With the exception of customers with backup sources, such as some larger commercial customers or dual feeds such as hospitals, the loss of a distribution feeder usually means that affected customers are without power. Moreover, the fact that all premises need to be served wherever they are located means that utilities must develop some creative and often ad hoc strategies to serve remote and economically disadvantaged communities, where the revenues often do not cover the cost of the capital improvements required to serve them, thus adversely impacting the utility's bottom line, and resilience often has to be sacrificed in order to support the capital investment required to serve marginal loads. The primary purpose of the electric distribution system is delivering reliable power to all premises every second of every day. Establishing or enhancing distribution automation means customers would experience fewer and shorter outages, reduced time required by field crews to restore power when outages occur, identifying weaknesses in distribution system components, and a safer and more secure work environment. In practice, this means pushing intelligence closer to the edge of the grid to not only report events back to a centralized location for analysis and further action, but also to execute localized and automated responses. Sophisticated applications running in conjunction with substation and distribution transformers can be designed to monitor activity and to adjust settings and even interrupt power in the event of a severe disruption.

Applications that are distributed closer to the edge of the grid are frequently run on robust platforms with redundancy and advanced cyber security controls built in. In many respects, they offer the ideal of greater efficiency, cost effectiveness, and security. However, the devil is always in the details. Like AMI (advanced metering infrastructure), this increased automation and centralized control means that inaccurate data and control by malicious individuals can magnify impacts when a vulnerability is exploited. Moreover, localized automation means that, just like smart meters, distribution applications now have many more entry points for hackers to wreak havoc. That means that any distribution solution risk management process must carefully consider the impacts associated with the processes it supports as the consequences of an attack could be much more catastrophic than simply a few houses losing power.

5.2 WHAT IS THE DISTRIBUTION SYSTEM?

The distribution networks of power delivery systems tend to be radial designs from substations out to customer locations with significant single points of failure. These radial designs provide little or no resiliency to equipment or power interruptions. Distribution automation systems deployment allows for more complex topologies to be built, including circuit ties and real-time analytics used to manage the complex topologies for power delivery. Advanced distribution networks allow for the integration of distributed energy resources (DERs), including generation and storage devices. These DERs can be leveraged in support of outage response, or in energy shortage periods, to manage supply in the distribution network for various periods of time. DERs also allow for greater flexibility in demand response programs by providing alternative energy sources local to consumption. DERs enable customers to manage their consumption and generation patterns. DERs will be operated in grid-connected and islanded modes. In the grid-connected mode, they will be utilized for peak reduction, loss minimization, and voltage/volt-ampere reactive (VAR) management. In the islanded mode, DERs will be used to supply energy directly to customer loads. The advanced distribution system will offer multiple new programs and capabilities, including

- *Two-way power flow capabilities:* Two-way power flow capabilities enable the distribution system to both supply power to loads and to leverage distributed power from plug-in hybrid electric vehicles (PHEVs) and consumer-owned generation sources. The advanced distribution technologies will need to be in place to achieve a true vehicle-to-grid system where PHEVs can be charged and also used as resources.

- *Load forecasting and predictive system modeling:* Load forecasting and predictive system modeling is based on ubiquitous data collection, so system planners will have a wealth of data to develop system planning models and outage modeling.
- *Risk analysis modeling:* Advanced distribution modeling allows for more complex topologies of distribution infrastructure to be modeled for risk and capacity challenges, and for expansion or construction to be enabled to manage those potential issues. Where simplistic radial topologies required less of this sort of modeling, new complex grid-style topologies will require more.
- *Real-time outage response:* Through the use of more complex topologies and DERs, the distribution system will be able to manage assets and load through real-time command-and-control systems, similar to evolving transmission systems. This includes responses to volt/VAR optimization and fault location, isolation, and restoration (FLISR).
- *Theft analysis:* Advanced distribution enables utilities to identify both power theft and asset theft through distributed measurements. Power theft can be automatically identified based on load and losses not matching supply values. Sensing and measurement allow utilities to pinpoint where this is occurring, allowing field personnel and authorities to track the issue. Asset theft can be identified and remediated without loss of service as systems can dynamically reconfigure themselves around the outage point.
- *Power quality improvements:* Customers have increasingly higher expectations for power quality as microelectronics expand their footprint. While all users expect a high-quality energy product, not all customers have equal requirements. Some commercial and industrial (C&I) customers may have significant requirements for very high levels of power quality in order to service fabrication, manufacturing, and other critical-use devices with very low power variance thresholds. Advanced distribution allows for improved power quality for all users and the capacity to deliver very-high levels of power quality to specialized users with specialized demands.
- *Dynamic feeder reconfiguration (DFR):* DFR will dynamically collect data from distribution feeders and, in case of a fault, will automatically isolate the fault and restore electric service using available capacity from adjacent feeders. The unique concept of this design is the real-time determination and transfer of available capacity from adjacent feeders. This is capacity that is normally not utilized with conventional (manual) load transfer schemes.

- *Dynamic and static islanding:* When upstream power delivery systems cannot deliver power, utilities can use "islanding" to manage generation and load in a self-contained area. A distribution feeder network may be an island if upstream power is cut off and local DERs provide power generation. This may be momentary or semi-long-lived. Distribution supervisory control and data acquisition (SCADA) systems would be used to open and close connections, as well as enable and disable DER power generation and delivery.
- *Advanced protection schemes:* Advanced protection will support availability of the system by executing automatic re-synchronization, relaying for loss of power or ground faults, reverse power protection, under/over-frequency, and voltage protection.
- *Peak-load management:* Advanced distribution works with AMI to both decrease peak-load through demand response (DR) and automatic load control (ALC), and to source supply locally from DERs in peak periods to provide transmission and generation congestion relief.
- *Micro-grid operation:* Micro-grids will contain several smaller power sources such as micro turbines, and photovoltaic and fuel cell systems. The micro-grid will appear to the larger electrical grid as any other regular load but it will be able to switch on and off the grid, depending on the price and availability of electricity. Micro-grids will ensure uninterrupted power to critical components.

5.3 DISTRIBUTION SYSTEM ARCHITECTURE

Distribution systems can vary significantly based on the technologies used; however, most solutions maintain the same functional areas:

- Utility field sensors
- Utility distribution and feeder meters
- Utility field controllers
- Local access network (LAN)
- Sensor/meter aggregator
- Wide area network (WAN)
- Data center access
- Sensor head-end
- Meter head-end
- Distribution SCADA master terminal unit (MTU)
- Traditional back-office applications

Figure 5.1 illustrates how these components interact with one another from a visual perspective.

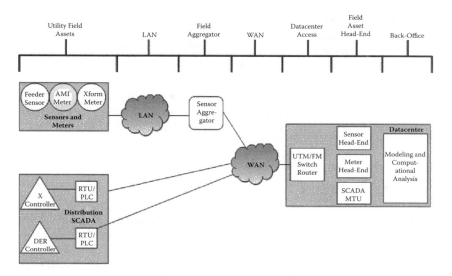

FIGURE 5.1   Distribution system overview.

5.3.1 Utility Field Sensors (Sensors)

Utility field sensors are used to actively measure power current status and quality at various points across the infrastructure. Sensors typically clamp onto medium- or low-voltage lines and inductively determine status and metrics of the power on the line. Sensors are passive measurement devices and typically have no capability to actively interrupt or impact services. Sensors are a fundamental unit of distribution automation as they provide measurement of specific locations on the distribution network. Information from sensors feeds both predictive and real-time system modeling and may inform processes that choose to dynamically reconfigure feeder or distribution system assets through SCADA systems. In this light, while sensors cannot directly impact power delivery, false, incomplete, or misleading data from a sensor could lead to "false positives" and inappropriate control system response.

5.3.2 Utility Distribution and Feeder Meters

Like its AMI cousin, the utility distribution meter is a solid-state metrology device owned and operated by the utility in order to measure and record utilization data at intermediate locations along the distribution network. Distribution meters may be placed with low-voltage transformers or other locations on the distribution network in order to identify power usage and trends. One key focus on distribution meters is identification of energy theft. By identifying "missing" power, especially in

dispersed rural areas, utilities can identify and isolate power theft locations. Feeder meters also assist in power quality measurement.

Utility meters are governed by American National Standards Institute
(ANSI) standards, much like customer meters, and likely integrate into
a common communications medium with AMI meters by leveraging a
LAN communication module for the same service. It is unclear at this
time if these meters would make use of connect/disconnect services. In
some models, they could function as circuit interrupters, but are more
likely to be passive measurement nodes similar to sensors.

5.3.3 Utility Field Controllers

Utility field controllers exist at the edge of the communication network, and
across the power delivery system. Field controllers are SCADA-type system control assets, including remote terminal units (RTUs), programmable
logic controllers (PLCs), and other communication device-enabled assets
that can control physical power system components such as actuators,
auto-reclosers, DERs, sectionalizers, etc. Utility field controllers provide for
remote system control and will vary from remote devices with localized
logic and remote operable reset to advanced system controllers that report
on state and action central system commands. Dynamic feeder reconfiguration and islanding are core computational processes that may command
field controllers to open and close circuits, and enable and disable DERs.

Field controllers are responsible for performing local automated
actions (such as opening a circuit when a voltage fault occurs), communicating fault data to master control stations, and executing reconfiguration commands from distributed control processes. As such, field
controllers represent a significant risk to infrastructure security as loss
of control, communications, or corruption of data could misinform distributed control measures and create widespread loss of service. With
the interconnection of feeder systems in support of DFR, impact areas
could be larger than just "a neighborhood."

RTUs and PLCs may use a wide range of communications mediums
to communicate to the master control system. Options include a wide
variety of public and private, wireline and wireless systems. Security
concerns may not approach transmission SCADA scale of impact, but
can be significant.

5.3.4 Local Access Network (LAN)

The distribution system LAN functions as a medium to aggregate meters,
sensors, and potentially other edge assets in a given region or area to a

meter/sensor aggregator. In many cases, the LAN for distribution automation will be exactly the same infrastructure used in AMI solutions in order to leverage investment in technology and deployment. In other situations, a utility may choose an alternate LAN medium technology for distribution automation edge assets. It is likely that different options would be considered for edge sensors/meters than would be for distribution automation controller assets. Distribution automation controller assets may end up using WAN services directly as opposed to using, or sharing, LAN access mediums with potentially lower security requirements.

There exists a wealth of LAN solutions in the marketplace and they span all manner of public and private, wireline and wireless technologies. Of the many solutions, they tend to fall into a couple of categories: industrial, scientific, and medical (ISM)-band wireless, private-band wireless, and forms of power line carrier. Each of these categories, and their specific communications technologies, has characteristics that change both the physical and logical solutions and either enhance or reduce security for the overall system. Please see the AMI Program Definition in Section 3.3 for an in-depth discussion of these technologies.

5.3.5 Sensor/Meter Aggregator

Sensor aggregators maintain the same characteristics and options as AMI meter aggregators do in terms of acting as field data aggregation and relay devices for edge data sources. Sensor or feeder meter deployments may or may not leverage the same technology, vendor, and assets as AMI meters, but the same options exist in terms of deployment models and technology options. In the end, sensor communications represent a similar problem statement as AMI meters, providing a reliable, but not necessarily mission-critical, medium for widespread two-way field data communications. In their application to islanding and DFR, AMI meters, feeder meters, and sensors all play a similar role in informing the computational modeling and control engines decision-making process.

5.3.6 Wide Area Network (WAN)

Wide area networks (WANs) provide a communication medium to interconnect the data center access and head-end systems to the field network devices. In most models, this is through the meter aggregator and the LAN, although some models connect directly via the WAN to the meter or edge node. WAN solutions can vary dramatically and range, like the LAN, across all manner of public and private, wireline and wireless technologies. Of the many solutions, they tend again to fall into a couple of broad categories: public wireless, private wireless, public wireline, and private wireline. Each of these categories, and their specific communications technologies, has characteristics that change both the physical and logical solutions and either enhance or reduce security for the overall system.

A key differentiator for distribution automation in comparison to AMI is the inclusion of mission-critical control systems across the WAN. In this respect, there is nothing that mandates that the AMI WAN, the distribution sensor WAN, and the distribution SCADA WAN be the same logically or physically. There are commercial pressures that make building separate services difficult, but there are also divergent security constraints that make building a single WAN difficult. This area highlights a key challenge for utilities looking forward in the IT and IT security domains. How do we meet stratified security constraints over common service areas that often have limited service offerings? This will be a key challenge to relate against cost, options, reliability, risk, and impact.

5.3.7 Data Center Access

The data center access component of the topology provides a secure bridge between the external network topologies and equipment, and the traditional secure zones within the data center where back-office applications operate. The data center access component may include any number of security devices, including firewalls, intrusion detection systems, and other security-layer assets. Additionally, it may terminate connectivity from multiple WAN services of differing capabilities and functions. It is expected that the data center access component represents the hub—or hubs in a high-availability data center model—for all manner of WAN functional services. These may include AMI WAN as well as other WAN components as additional components for distribution and transmission.

5.3.8 Sensor Head-End

Sensors act in a similar fashion to meters when it comes to head-end software. The head-end software acts as an application-layer interface to the field sensors. The head-end presents sensor data to other back-end applications via standards-based interfaces such as service-oriented architectures (SOAs) using eXtensible Markup Language (XML)-based protocols as well as managing configuration (possibly including Geographical Information System [GIS] and Global Positioning System [GPS] data), firmware, and device status. It is likely that there is no other mechanism for interacting with or collecting data from the field sensors than through the head-end platform. Sensors are not currently defined (as far as I can see) through ANSI-style standards, and therefore may be Internet Protocol version 4 (IPv4), IPv6, or something else entirely. Sensor head-end software runs on standard Windows® (Microsoft Corporation) or UNIX® (X/Open Company) operating systems and uses standard database software for data storage.

5.3.9 Meter Head-End

Each manufacturer of metering communications modules offers a specific set of management applications used to communicate between the data center and the meter communications modules in order to collect data, manage the device configuration and firmware, and execute potential internal controls such as connect/disconnect. The head-end application software is highly specialized control software and should be treated as such in the security review. In many cases, these head-ends will exist in the DMZ (demilitarized zone) and may be deployed outside the typical data center for the utility, thus opening up additional security questions.

Distribution feeder meters, and potentially sensors, may be controlled through a common meter head-end application, or through an alternate, but comparable, meter head-end platform. While meter head-end platforms tend to be tied to the LAN communications vendor today, that should become less and less so as time goes by and ANSI C12.22 becomes more established. Standards for communications services should allow for these to be vendor-interoperable as time goes on.

5.3.10 Distribution SCADA MTU

Distribution system SCADA systems will have a master control station where control actions, either manual or automated, are implemented and communicated out into the distributed control system. It is unclear if the SCADA MTU system in the distribution network will be a physically and logically disparate system from the transmission SCADA MTU, or if it will be integrated into the existing transmission SCADA system. Either way, the master terminal will be the secure implementation point for DFR, islanding, and an all-dynamic distribution reconfiguration.

5.3.11 Back-Office Computational Platforms

Distribution and wider Smart Grid evolutionary capabilities depend on complex computational modeling and analytics software to identify system inefficiencies in steady-state or outage conditions, and to respond to outage or inefficient operating conditions. The computational platforms will evaluate sensor and meter measurements, along with distribution SCADA system stats and traps, to identify issues and opportunities, and either recommend or act upon system topology information to respond. These platforms operate within the core of the data center back-office in secure zones, but are dependent on real-time information from a wealth of sources across a service territory.

5.3.12 Traditional Back-Office Applications

The traditional back-office consists of various hardware and software technologies used in processing the information received from all the dimensions of the utility environment. These include billing, outage management, GIS, and work management applications.

5.4 DEFINITION OF DISTRIBUTION AUTOMATION

Distribution touches nearly all aspects of the electricity delivery system—from the consumer to marketing and operations. It is one reason that regulators are reevaluating whether distribution should continue to be exempt from North American Electric Reliability Corporation Critical Infrastructure Protection (NERC CIP) cyber security requirements. In any case, the interfaces between distribution components must be reviewed carefully for cyber security risks, and appropriate controls should be implemented. Let's now take a look at the individual components and their relationships to other functions.

At its most basic level, distribution has changed little in the hundred or so years that electricity has been commonplace. Electricity is generated and sent over transmission wires often running hundreds or perhaps thousands of miles at high voltages up to 765,000 volts. Once the transmission lines reach a local area needing electricity, the voltages are stepped down to a more manageable level between 2,400 and 34,500 volts, but usually in the 13,000-volt class. This occurs at distribution substations that also act as key control points for ensuring voltage and quality levels meet the requirements of the end loads, and as one of the primary areas where distribution automation is being targeted. From the substation, feeder lines serve as the backbone of the distribution system, bringing electricity to local neighborhoods. These feeder lines can be the more common radial lines or a loop configuration that provides greater redundancy in the event of failure. A third configuration, known as a primary network system, is comprised of a series of interconnected

FIGURE 5.2   Secondary transformer. (Reprinted from John D. McDonald *Electronic Power Substations Engineering, Second Edition.* ©2007. New York: Taylor & Francis Group, LLC. With permission.)

primary feeder lines from various substations and is intended for densely populated areas, but is currently in limited use. Connecting to the primary feeder are secondary systems (see Figure 5.2) that provide service to individual customers through a series of transformers connected to primary feeders that step down voltage once again to levels we're accustomed to using in our homes and businesses.

These include single-phase systems for residential customers providing service up to 240 volts and three-phase systems for commercial and industrial customers with levels up to 480 volts. It is really at this point where the automated metering functions described in Chapter 3 take over. However, like all parts of the grid, the lines are blurred as many customer-based meters are also used to measure voltage levels and influence other decisions made for distribution automation. For example, a decision to shed load by sending signals through the AMI system directly affects the operations of the distribution management system. Additionally, customer-based generation, or distributed generation, heavily relies on signaling provided by AMI systems, but it is intricately related to the existing distribution system. Given its significance, we discuss distributed generation in Chapter 7.

For all these configurations, various sensors and actuators are deployed throughout these feeder lines to protect the grid from damage, limit outages, and allow for the rerouting of electricity. A sensor provides information locally or to remote applications about the state of electricity flowing through a line or the condition of a component such as a capacitor, transformer, or circuit breaker. They may measure voltage level, current, harmonics, heat, gas levels, and other conditions. Actuators are a class of devices that take an action automatically or based on a control signal based on particular events. This could include a circuit breaker or fuse that trips when there is a disruption in the line. They can also be used to reenergize a line after the disruption has passed or a repair has been made. In many cases, devices contain both sensor and actuator functionality. For example, a circuit breaker detects a spike in the current that is outside permissible limits and opens the circuit. An automatic recloser has similar capabilities but can also automatically reenergize the line. As the distribution system gets smarter, these local and somewhat simple devices are getting more sophisticated, with computer software being used to perform complex calculations at local and remote locations in order to further optimize voltage levels, reduce frequency fluctuations, and reduce outages. Consequently, the reliability of the data inputs, the algorithms for processing the data, and the resulting control signals will all be of greater importance, thereby increasing the importance of cyber security controls.

The notion of a Smart Grid is really the idea of integrating enhanced communications technologies with enhanced decision making and automation of actions, both centrally and in the field. Consequently, Smart Grid doesn't fundamentally change the functions of a transformer, capacitor bank, circuit breaker, or switch. Instead, the technologies added allow for better monitoring and control of the state of these devices. For example, several vendors market products that measure the health of a transformer using a variety of methods, including gas chromatography, temperature, and usage patterns, to determine when a failure is likely to occur. Given that many transformers are still in operation beyond their life expectancy, it certainly helps to show why such analysis would be useful. It also demonstrates the potential damage that could result if utilities rely on data that have been intentionally or inadvertent manipulated. It can take as long as 18 months to replace some of the larger, more expensive transformers. Applying appropriate physical and cyber security protections is essential in those circumstances to not only allow the utility to predict failure, but also to ensure that devices are not abused to the point where their life spans are shortened. For automatic reclosers and the circuit breakers they control, safety is also a concern. More and more, these devices require information technology to operate correctly, for example, SCADA systems. These can trip circuit breakers and note when the circuit is open or closed. A maintenance worker may actually have to call the SCADA operator to find out if a line has been de-energized before performing work as visual inspection may not be reliable. One can clearly see the importance of that information being correct and the potential harm that can result from unauthorized manipulation of that information.

However, understanding the security risks with distribution automation goes beyond the state of a device and its representation on a computer screen. These so-called human-machine interfaces (HMIs) have been around for decades in various forms. The kinds of automation that Smart Grid is imposing add additional layers of complexity that are more than simply a one-for-one representation of the device state, which is essentially what an HMI provided. Instead, modern distribution management applications take in information from hundreds, thousands, or potentially millions of devices; aggregate that data; and then either advise or automatically take an action based on an analysis of that data. Much of the interaction happens in the domain called Operations, which is where new software applications are deployed to take in massive amounts of information and provide guidance on everything from billing rates to the voltage levels coming from a generation plant.

Unfortunately, many of these security risks are not reflected in the same kinds of guidance and regulation that typically are applied to transmission. Because distribution is not perceived to affect the stability of the bulk electric system, the requirements of NERC CIP discussed in Chapter 2 normally do not apply. However, simply applying NERC CIP to the distribution system is not that simple. Because of the many different assets involved in the distribution system, it is difficult to determine which ones would be critical. Applying the current state of NERC CIP to the distribution system will likely be an impossible task. New criteria must be established in order to apply the standards to the distribution system. Furthermore, there is no regulatory authority with the ability to regulate the distribution system. But if there were, how would assets be cataloged for impact, and what would be the scope of the impact? More than likely, the methodology would include criteria that represent critical community resources, such as airports, hospitals, military installations, etc. A critical asset in distribution would then be the control center that provides supervisory and data acquisition for the system as well as distribution assets that provide electricity to AMI resources that support critical facilities. For example, what would the impact be if the distribution system could be accessed and used to shut off power to a hospital? From a financial perspective, what would be the impact if consumers were not able to sell power back into the grid?

Moreover, in such a case, we would be talking about the distribution's effect on society versus the bulk electric system. That is because distribution assets that affect the bulk electric system are already covered by NERC CIP. As a result, new criteria must be developed or a new asset to protect must be identified. The likely scenario is that critical infrastructure as defined by the Department of Homeland Security

(DHS) in Homeland Security Presidential Directive 7 (HSPD-7) will be the sectors that are supplied energy from the distribution system that is considered critical. The following represent critical infrastructures as identified by the DHS.[1] These agencies all maintain regulatory capabilities of some sort and potentially could be used to regulate energy that is provided to them.

- Department of Agriculture: agriculture, food (meat, poultry, egg products)
- Health and Human Services: public health, healthcare, and food (other than meat, poultry, egg products)
- Environmental Protection Agency: drinking water and water treatment systems
- Department of Energy: energy, including the production refining, storage, and distribution of oil and gas, and electric power, except for commercial nuclear power facilities
- Department of the Treasury: banking and finance
- Department of the Interior: national monuments and icons
- Department of Defense: defense industrial base

That is, the risk-based methodology used to declare a distribution system asset as critical would require that the impact be based on criteria such as what is established in HSPD-7. For example, what is the impact, for example, if power was lost to the water system and, as a result, drinking water could not be produced? Components that make up the distribution system would then be related to the national and economic security of the United States.

Before delving into the various use cases that make up distribution automation, let us examine more closely the elements of distribution automation. RTUs play a vital role in facilitating communication between centralized applications and devices in the field. These are devices with embedded microprocessors that control the physical operation of a grid device such as a switch or an auto-recloser. While the definitions of these devices vary and overlap, an RTU is typically considered the part that handles remote communications, typically over protocols such as Distributed Network Protocol version 3 (DNP3) and MODBUS® (Schneider Automation, Inc.), which may be over Ethernet, serial, or other media. Regardless, the communications and on-board intelligence associated with these modern devices means that lines can be de-energized and energized with no human interaction, and they can also be re-programmed and controlled remotely. Consequently, security is important. The RTU is part of a larger set known as intelligent electronic devices (IEDs), which are typically microprocessor-based devices that are programmable and capable of controlling a variety of end devices. IEDs are used to control auto-reclosers, circuit breakers, capacitor banks, voltage regulators, and other devices.

Traditionally, these devices have been located in substations, many of which were only accessible via dial-up modem. That limited the ability to monitor the devices remotely in real-time and meant that many of them were at least theoretically accessible by anyone with a telephone and a modem. However, in practice, many substations using dial-up take advantage of various secure dial-up software products that control access to substation devices. In some ways, dial-up inhibits attackers accustomed to scanning thousands of devices at once over the Internet and requires the less-efficient war-dialing method. However, as network connectivity has improved for most utilities, dial-up is being replaced by other connectivity, for example, fiber-optic cable runs between the utility offices and substations, microwave, and wireless technologies such as Worldwide Interoperability for Microwave Access (WiMax). Nonetheless, legacy support requirements mean that older protocols such as DNP3 and MODBUS are run over these new transport media. These protocols are known for their poor security, which means that even if the end device requires adequate authentication, an attacker may be able to impersonate a legitimate user. Consequently, it is important that these protocols be supplemented with more robust security options such as encrypted tunnels for wireless and fiber dedicated for the utility alone. While the latter option is usually feasible where fiber is used, utilities are frequently seeing multiple grid functions competing for that same fiber network. For instance, individuals responsible for AMI are seeking to piggyback on top of the substation-to-plant networks in order to provide the WAN backhaul from collector to head-end rather than pay the cost of dedicated fiber or use a carrier network that is less secure. Protocols such as IEC 61850 (Communications Networks and Systems for Power Utility Automation) are seen as more robust replacements for DNP3[2] and MODBUS[3] in the area of substation communications.[4] The protocol has built-in security functionality that DNP3 and MODBUS lack. However, utilities will still need to support legacy protocols in the near term, and replacement of devices tends to happen slowly unless additional functionality is required. Until then, utilities need to ensure that adequate security is implemented to compensate for the weakness in these protocols.

GISs are also part of the core capabilities of a distribution system. When combined with an outage management system (OMS), they provide the utility with a way to quickly detect and locate outage and system faults. From a cyber security perspective, the GIS is probably not as critical. However, the fact that its information is often drawn from outside sources means that its reliability might be in question. Nonetheless, the impact is usually delayed as alternate resources such as online and paper maps are also available to supplement what the GIS offers.

Distributed intelligence capabilities can be the traditional IEDs, which are a class of relatively simple devices that control breakers, capacitor banks, and various other electromechanical devices connected to the electrical grid. These may include RTUs and PLCs. However, newer devices are expanding the definition of "intelligent." More often, these embedded devices are running robust operating systems that can analyze massive amounts of data and take actions.[5] This can be both positive and negative. On the positive side, increased intelligence allows utilities to quickly identify a problem at its source and take action at a local level. This then allows for isolation of the problem without disrupting a significant portion of the electrical grid. It can also detect and report on potential security events at their source, resolving one of the electrical grid's most vexing challenges, that of quickly locating the source of a problem. On the negative side, putting so much processing capability at so many remote locations means that a compromise of one of these devices could give the attacker the ability to leverage that access to not only wreak havoc locally, but also to work his way up to the centralized control systems by relying on trust relationships. An effective security design that limits how much trust is accorded to a single node helps to mitigate much of this threat.

Notwithstanding the wide variety of field devices placed along the electrical grid, the key control points at the distribution level reside at the substation. For many grid networks, the distribution substation is currently the last place the utility can remotely control the flow of power to the customer. While AMI technology and home area networks (HANs) promise to give more granular control, the instrumentation and control functions at those levels are likely more basic with fewer features than what a SCADA system is able to control at a substation.

5.5 HOW DOES DISTRIBUTION AUTOMATION WORK?

"The supervisory control functions of electric utility substation automation (SA) systems provide routine and emergency switching and operating capability for station equipment. SA controls are most often provided for circuit breakers, reclosers, and switchers. It is not uncommon to also include control for voltage regulators, tap-changing transformers, motor-operated disconnects, valves, or even peaking units through an SA system."[6] Figure 5.3 highlights the breakdown of SA functions.

There is a wide variation in the types of communications technology used and how it can be implemented. For older implementations, the actual power equipment may be a stand-alone device with its input limited to a serial or 4- to 20-milliamp circuit that received the most basic

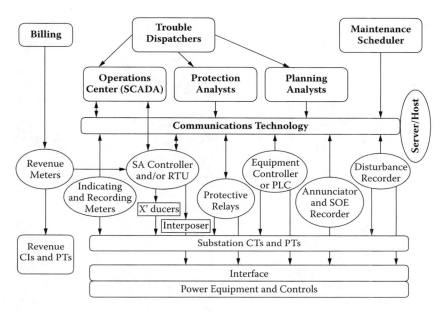

FIGURE 5.3   Distribution automation function. (John D. McDonald. 2007. *Electric Power Substation Engineering*, CRC Press, Inc., Boca Raton, FL.)

of analog signaling. For example, a recloser may simply have a write capability to change its state from open to closed and vice versa, and a read capability for engineers to query the state. Newer devices not only provide greater intelligence and instrumentation capabilities, but also integrate modern communications interfaces such as Ethernet and often have their own built-in authentication and access control features. In the past, substation automation design required exhaustive examination of the device requirements, their location in the substation, the availability of controller devices such as IEDs, RTUs, and PLCs to control the power equipment, and the nature of the work performed in the substation (see Figure 5.4).

With newer technology, providing connectivity between a recloser or protective relay and the SCADA system may be no more difficult than adding a computer to a corporate network. Nonetheless, an end device that has a built-in Ethernet card, and perhaps even a built-in Web service, may still use protocols or messaging standards that are foreign to the SCADA system. It is often the case that a device driver for a SCADA or distribution automation system must be developed. This is also where security concerns are raised. Any device driver is effectively providing a translation between the protocols, file formats, and control messages used by the device and those used by the control system meant to talk to

FIGURE 5.4   Utility substation.

it. And just like human language, translating instructions across varied devices is bound to result in a few situations where one-to-one mappings do not exist. For those concerned with security and performance, it is critical that a common understanding is formed. Many manufacturers intentionally develop proprietary protocols and protocol extensions to make it more difficult for customers to implement multi-vendor solutions that could limit the manufacturer's revenue potential. Even among commonly used protocols such as DNP3 and MODBUS, there are a variety of implementations that communicate on different ports and use extensions to the protocols that are not commonly used. From a security perspective, it is important to not only understand how to interact with a new device to perform the required function, but it also is necessary to understand any other functionality that the device may expose. This can be accomplished by reviewing the vendor's detailed specifications for the protocols and services implemented, independent security testing, or internal product testing. In many cases, where DNP3, MODBUS, or another legacy protocol is used, authentication and access control will be provided by some other device such as a virtual private network (VPN), remote access server, or a physically protected point-to-point connection.

Additionally, it is important to recognize the significance of the activities outside the substation. While SA incorporates some intelligence locally (one of the driving themes of distribution automation is to auto-mate some routine substation events locally), it also empowers engineers to act remotely so substation visits are sharply reduced and the util-ity has the benefit of looking at information from multiple substations, pole-top transformers, and line monitors all at once. That allows engi-neers and operators to quickly identify problems and prevent them from cascading. It also facilitates further optimization of the grid by measur-ing voltage drops at various locations and calculating the level of genera-tion needed to sustain the load. The *Security Profile for Distribution Management* developed by The Advanced Security Acceleration Project for Smart Grid (ASAP-SG) defines four networks covered by distribu-tion management (DM). They include the DM Field Network, which includes the substation and field devices we've been discussing; the DM Control Systems Server Network, which includes the servers and other network devices centrally located at a utility data center that directly communicate with the field network and its devices; the DM Control Systems User Network, which includes the individual user worksta-tions that send control instructions and view status information; and the Non-DM Utility Network, which is effectively everything else in the utility's infrastructure.[7] For typical distribution-only utilities, the first three networks have collectively been known as the SCADA network or process control network and were often considered a single network that was physically isolated from the rest of the utility. However, with Smart Grid, that isolation is no longer feasible. Data from the DM Control Systems Server Network must talk with AMI networks to determine voltage levels at the customer meters or initiate a load control event. Various other data must be fed to outage management systems that need to be accessible by customer service representatives using the traditional enterprise network.

It is this interaction between multiple applications that raises secu-rity concerns. In the past, a SCADA system was purchased off-the-shelf from a single vendor who then helped integrate it with the various field devices. In many cases, the vendor actually designed the network. In con-trast, a demand response application, which takes in a vast array of mea-surement readings from various systems and determines everything from what price signals to send out to which feeder lines might qualify for a rolling blackout, is generally a new application from a vendor different from the SCADA system vendor. Similarly, the Meter Data Management System that may feed both the SCADA system and the demand response application comes from yet another vendor. While most of these applica-tions are leveraging common TCP/IP protocols with Web services and SQL (Structure Query Language)-based database queries, the fact that these require utilities to integrate them together or hire someone to do it for them is something new to many utilities accustomed to buying a system from a single vendor that was fairly self-contained. In fact, third-party applications like anti-virus and log management are offered bundled by a vendor to make the process more seamless. Smart Grid is forcing many utilities to become large software development shops as more of the focus moves from linemen climbing poles to inspect, fix, or replace equipment to IT professionals responsible for managing complex algorithms that can detect problems and often come up with automated fixes. This means it is imperative that the right data are collected and processed. While remote control of field devices has existed for some time now, it has never been more critical that the aggregate data coming from those devices are correct. Sending a system bad information is often the equivalent of hacking the system directly and executing the wrong control instructions. The results are often the same.

5.6 DISTRIBUTION SYSTEM COSTS

The costs associated with the distribution system are considered significant because of the sheer number of assets associated with the system. There are numerous distribution substations that are needed to get data from the transmission system and eventually the distributed generation system to the feeders that feed energy to households. The cost can, for the most part, be calculated by the number of customers that exist within a given service area. For example, if a distribution substation serves 10,000 customers, then over 300 substations will be needed to support a customer base of 3 million customers. This is not to mention all the components that exist within these stations, including the transformers on poles and various sensors that must exist throughout the distribution system network. This example is meant to illustrate the amount of equipment involved in supporting the distribution system. In terms of reliability losses, these are limited because of the distributed nature of the system.

5.7 WHAT IS THE SMART GRID FUNCTION OF DISTRIBUTION AUTOMATION?

While substation automation is an innovation that predates so-called Smart Grid applications, it is a critical component of any distribution management application because it facilitates the automated controls needed to make distribution management possible, such as the IEDs controlling capacitor banks to maintain voltage levels or the reclosers that effectuate automated feeder reconfigurations. The *Security Profile for Distribution Management* calls out the following centralized applications as being part of a distribution management system:

- "Distribution SCADA applications
- Volt/VAR control
- Fault location/isolation, and service restoration
- Automatic feeder reconfiguration"[8]

The *distribution SCADA application* is the core of many of the traditional distribution automation applications. While it typically has not included customer load control capabilities, it does act as the primary application to control substation and distribution network devices. It monitors the capacitor banks and reclosers and may also incorporate a variety of other sensor data.

Additionally, future applications will be designed to measure transformer performance and other pole-top uses. A traditional use of this distribution SCADA application has been fault location, isolation, and restoration. Typically, this involved the distribution SCADA system being used to communicate with circuit breakers and reclosers to monitor their status and respond to faults.

Among the new features being introduced under the Smart Grid banner is *volt/VAR control*. "The distribution-level power system may experience over-voltage/under-voltage condition that can be mitigated using volt/VAR control (VVC). The objective of VVC is to minimize the power loss while maintaining acceptable voltage levels and distribution substation power factor limit (reactive power limit). VVC optimization is typically made possible by controlling the tap position or by varying the shunt capacitance. To regulate the voltage output on a distribution transformer load tap changer (LTC) or regulator is used. Switchable capacitor banks are used to provide the reactive power compensation. VVC is a central application that receives the voltage levels at each bus and reactive power requirements in the distribution network. The central application computes the power flow in the network under given constraints and communicates the set points to LTC and capacitor banks. While LTC executes the set point from the central application, the switchable capacitor bank controller has a field application that will discretize the solved capacitance set point to control the banks of capacitance."[9]

Another feature is *automatic feeder configuration*. "Automatic feeder reconfiguration is a distribution operations planning application used for closing and opening switches within the distribution network system (whole system or groups of sub-systems) to restore power to portions of the network after contingencies and topology changes (varying loading conditions). This application of pattern describes a fully distributed mechanism for feeder reconfiguration. The feeders in the distribution system are equipped with intelligent electronic devices (IED) which are wireless automatic reclosers capable of forming a mesh network to autonomously communicate with each other without involvement of a central application. These IEDs can locally sense faults within a sub-system and communicate the status to the neighboring feeders. The feeders then react intelligently to reconfigure the distribution network topology to restore/de-energize service to sub-systems."[10]

5.8 THE IMPORTANCE OF THE DISTRIBUTION SYSTEM AND ITS SECURITY CHALLENGES

The current distribution system, without distributed resources and without an intelligent networked configuration, limits its ability to be resilient. Linear radial distribution networks have little capability for automated outage identification, isolation, and response. This creates a significant challenge in delivering advanced power capabilities to customers.

The operations and outage response model of the distribution network has remained the same for years in most utilities. Current utilities identify distribution-level and customer issues primarily based on customer calls. Equally, outage resolution depends on customer contact and acknowledgment. With Smart Grid distribution systems, outages are identified and located in real-time prior to customer contacts. This allows rapid deployment of resources to the right location to resolve issues. In distribution automation programs, distributed generation and automated switching, self-healing topologies may mask the issue from customers while the issue is resolved by the utility. In this way, distribution automation transforms the distribution network into a high-availability power delivery system capable of meeting the key characteristics of the Modern Grid. Additionally, detailed availability and quality metrics are available for tracking customer service quality.

Distribution automation is being developed and deployed to service the following key goals:

- Improve reliability and quality
- Improve security and security response
- Improve environmental impact and public safety
- Improve system outage identification and response
- Improved response to system requirements change
- Defer business-as-usual capital investments for upgrades
- Integration of PHEV load
- Increased penetration of distributed energy resources

The primary goal of distribution automation is to improve power delivery system reliability, performance, and quality. By collecting, analyzing, and actioning comprehensive sensing and measurement data, distribution communication networks can be built and operated at much higher levels of availability and quality over current models. The distribution network of today has numerous and visible single points of failure, making service disruption due to cyber or physical attack a serious risk. Distribution automation allows for distributed energy sources and dynamic reconfiguration, making resilience to targeted or coordinated attack far less likely to interrupt or impact power services. Distribution automation provides for significant public safety improvements both by improving service quality and availability for key life-supporting technologies, and also through predictive analysis of systems to predict and prevent potentially hazardous conditions. System losses will also be reduced through the implementation of AMI and distribution automation, decreasing generation demands, and, in turn, reducing emissions. Distribution automation will use as its core a comprehensive measurement and data collection system. These data, and associated modeling components, will allow utilities to rapidly identify and remediate outage or quality issues. As areas were built out and requirements grew, historical responses to system demand were based on new feeders and "express circuits" to meet demand. These solutions can be difficult, costly, and time consuming to get approved and commissioned. Distribution automation allows for existing infrastructure to be further leveraged, speeding deployment and response and to defer or avoid some planned capital investments by improving the utilization of the existing system. It also allows utilities to extend the lifetime of the power equipment by being exposed to fewer short-circuit currents. Without a distribution automation system, the charging needs of PHEVs at scale would overpower the current electrical grid. Advanced Distribution Optimization (ADO) would allow vehicles to charge their packs during the most beneficial periods to the consumers and utilities. The concept of vehicle-to-grid may also be realized and allow vehicles to provide power back to the grid during peak periods, if necessary. This would lead to increased reliability, reduction in emissions, and more stable prices in electricity. Utilities that have more distributed energy resources or distributed generation (DG) at their service can carry less generation in reserve. This would, in turn, result in lower emissions. ADO would also permit faster demand response, thereby improving overall reliability.

5.9 SECURING THE DISTRIBUTION SYSTEM

As with any power system asset, cyber security should ensure that it considers the functionality of the system and the need for reliability before cyber security principles can be applied. The advanced distribution system is extremely complex, with a variety of moving parts and components that will leverage IT systems for automation. There is danger from this aspect in that many things can happen that will require advanced monitoring and prevention. Smart security, in other words, will be needed for Smart Grid technology. In the past, intrusion detection systems, log monitoring, and access control processes and procedures have been used in separate instances in order to provide protection. More recently, however, event correlation has become an increasingly popular way to detect and prevent advanced threats. It is this correlation technology that will have to be much stronger to prevent threats from causing problems in the distribution system, among others. Cyber security defenses will likely need to correlate events on the power system to events that happen on the information technology assets that provide automation. For example, if a malicious payload is detected on the intrusion detection system that targets a known vulnerability on the controller, we would probably understand that a cyber-threat had just compromised the controller in question. This capability, however, currently does not exist.

To achieve this level of correlation, information technology cyber security and the power industry are going to have to work together in order to develop a solution that will meet all parties' objectives. These objectives, however, first must be defined and clearly articulated to one another. In the case of information technologies, they will want to enable grid automation. While cyber security will want to develop mechanisms to protect grid automation, the power industry is going to desire grid automation. From these high-level objectives, we can start to accomplish more low-level needs. The power industry will identify what it wants; the IT industry will develop solutions to achieve those goals; and cyber security should look to determine what could go wrong based on what the power industry is asking for and what IT is providing.

Traditional enterprise IT systems are used for data communication but generally do not maintain critical physical infrastructure. When enterprise computers crash, data can be lost, but things generally do not explode. When data communications links fail, humans can often find alternative methods for communicating. Smart Grid machine-to-machine systems control mission-critical infrastructure where failures can have widespread impact. As a result, cyber security in IT for power system assets must be clearly understood in order to provide adequate protection.

5.10 DISTRIBUTION MANAGEMENT SYSTEMS

While the field and substation devices described above certainly contribute to greater distribution automation, they are only part of what a smarter grid is all about. Ultimately, distribution is about making better decisions and automating the process. This is where a distribution management system (DMS) proves valuable. It represents the back-end applications referenced earlier. A DMS is typically a series of software applications running on Windows- or Linux-based servers located at a utility's data center that interact with various systems, including outage management systems, geographic information systems, AMI head-end servers, SCADA systems, customer information systems, and other systems responsible for controlling or monitoring the components that make up distribution at a utility. Principally, the four areas a DMS covers are field devices (e.g., remotely controlled switches, relays, other IEDs), communication systems (e.g., radio, fiber optics, WAN and LAN technology, RTUs), control and dispatching center (e.g., SCADA, OMS, GIS), and integration with enterprise applications (e.g., enterprise resource planning (ERP) platforms, customer information systems, human resource systems).

A DMS fundamentally provides centralized automation of distribution resources. However, more significant may be its decision support role. While distribution automation has existed in a variety of forms for decades, the sophistication offered in the decision support area is somewhat new. The algorithms used can be quite complex and rely on neural networks and other artificial intelligence concepts. What that complexity means is that while a human may be in a position to intervene before an action is taken to reduce or increase voltage levels, the basis for that action may not be transparent. Instead, he or she must simply trust that the data received are correct and the algorithms were designed correctly. This makes it incredibly important that cyber security be protected, not with just the control functions, but in each and every instance where data are collected. Additionally, these applications must be tested and their code evaluated for security vulnerabilities, as SOA-based applications are notorious for their security bugs as each application can be assembled separately and introduced into an architecture that was never initially envisioned. This substantially reduces costs and expands the usefulness of the application, but not without trade-offs. However, using sound security integration and review practices, the risk can often be brought to acceptable levels.

5.11 STANDARDS, INOPERABILITY, AND CYBER SECURITY

Not surprisingly, the Smart Grid revolution has spawned a lot of standards drafting efforts by many of the standards development organizations (SDOs). Among them have been the efforts of Technical Committee (TC) 57 of the International Electrotechnical Commission (IEC). That group has focused on topics related to distribution. One of the challenges at this point is the role standards should play in a rapidly evolving industry, particularly in the realm of cyber security. On January 31, 2011, the Federal Energy Regulatory Commission (FERC) held a technical conference to consider a proposal to adopt five standards developed by the TC 57.[11] While it was unclear what adoption really meant, the consensus of those testifying was that there was not sufficient consensus in the industry for any level of adoption, particularly in the area of cyber security where some of the guidance was incomplete, outdated, or just wrong. The reason we are discussing this here is to convey the challenges faced by individuals seeking to provide more than high-level cyber security guidance, such as is included in NERC CIP. Many of the TC 57 standards go to great lengths to describe interactions between components of a distribution system and describe specifically the data exchange formats. However, cyber security is only covered in one of the documents, IEC 62351,[12] and even then the coverage is limited. Older ones such as IEC 60870 are being brought to the forefront to address the increased needs for automation and coordination between control centers. "The Telecontrol Application Service Element (TASE.2) protocol (also known as the Inter-Control Centre Communications Protocol, ICCP) allows for data exchange over WANs between a utility control center and other control centers, other utilities, power pools, regional control centers, and non-utility generators. Data exchange information consists of real-time and historical power system monitoring and control data, including measured values, scheduling data, energy accounting data, and operator messages. This data exchange occurs between one control center's supervisory control and data acquisition/energy management system/distribution management system (SCADA/EMS/DMS) host and another center's host, often through one or more intervening communications processors."[13] The original ICCP described in this standard has limited security, with authentication dependent upon security through obscurity. Essentially, there is a somewhat obscure table structure that must be navigated to obtain and modify information transmitted between control centers.

A newer standard, IEC 61850, as noted above, is designed to replace some legacy protocols such as DNP3 (Distributed Network Protocol 3) and integrate with others such as ICCP. "IEC 61850 addresses the data exchange on three levels: process level, field level, and station level. It defines the following four important aspects on these levels: standardized self-describing data, standardized services, standardized networks, and standardized configuration for a complete description of a device."[14] Currently, the standard is used for substation communication, such as between an IED (intelligent electronic device) and a controller. IEC 62351, part 3-6, addresses cyber security for IEC 61850. "Here, IEC 62351 specifies cipher suites (the allowed combination of authentication, integrity protection, and encryption algorithms) and also states requirements to the certificates to be used with TLS (Transport Layer Security)."[15] Authentication is also leveraged through this process.

By contrast, IEC 61968 focuses on the distribution management system described above, tying together the various subcomponents of distribution, such as substation automation. "As used in the IEC 61968 series, a DMS consists of various distributed application components for the utility to manage electrical distribution networks. These capabilities include monitoring and control of equipment for power delivery, management processes to ensure system reliability, voltage management, demand-side management, outage management, work management, automated mapping and facilities management."[16] IEC 61970 provides similar guidance at the application programming interface (API) level for energy management systems.[17] Both IEC 61968 and IEC 61970 comprise a common information model (CIM) for utility data exchanges and application interoperation that can be applied to several different functions. For IEC 61970, "[t]he common information model (CIM) specifies the semantics for this API. The component interface specifications (CIS), which are contained in other parts of the IEC 61970 standards, specify the content of the messages exchanged."[18] However, there is currently limited guidance with respect to cyber security in either of those standards or in IEC 62351, which is responsible for cyber security for the TC 57 family of standards.

In addition to the TC 57 family of standards, in the particular the CIM standard, an alternate scheme for distribution called MultiSpeak® has been sponsored by the National Rural Electric Cooperative Association (NRECA). That standard is targeted at smaller utilities that have fewer resources. "Hence, MultiSpeak has focused from the first on the development of tightly specified profiles of data objects and standardized implementations that vendors could install, largely unchanged, at many utility sites."[19] Nonetheless, the two standards perform similar functions. Like the IEC standards, much of security in the MultiSpeak standard is left to vendor implementation as it uses standard service-oriented architecture (SOA) protocols such as Simple Object Access Protocol (SOAP) and Web services that implement security at a layer below that of MultiSpeak.

With all these standards, it is clear that complying with them will not be sufficient to demonstrate adequate security. It will be up to control frameworks such as NERC CIP and various security best practices to be leveraged. There are currently a number of nascent efforts focused on more detailed conformance cyber security criteria and certification regimes for Smart Grid applications and equipment. Ultimately, it will be up to the industry, and perhaps the regulators, to ensure that adequate cyber security measures are defined and that various products can be evaluated.[20]

ENDNOTES

1. For more information about the sectors deemed critical by HSPD-7 and related issues, see http://www.fas.org/irp/offdocs/nspd/hspd-7. html.

2. "Distributed Network Protocol Version 3 (DNP3) is used by SCADA systems to communicate between the Master host and the Slave units. This infrastructure is open, and effective authentication or encryption mechanisms do not exist. Although the utilities have increased their attention on improving the security and reliability of the SCADA systems in recent years, many owners and operators do not yet have the technology, tools, capabilities, and/or resources needed to secure their systems." Munir Majdalawieh, *DNPSec: Distributed Network Protocol Version 3 (DNP3) Security Framework*, December 19, 2005, p. 2 (found at ttp://www.acsac.org/2005/techblitz/majdala-wieh.pdf).

3. "MODBUS is an application-layer messaging protocol, positioned at level 7 of the OSI model. It provides client/server communication between devices connected on different types of buses or networks. The de facto industrial serial standard since 1979, MODBUS continues to enable millions of automation devices to communicate. Today, support for the simple and elegant structure of MODBUS continues to grow. The Internet community can access MODBUS at a reserved system port 502 on the TCP/IP stack. MODBUS is a request/reply protocol and offers services specified by function codes. MODBUS function codes are elements of MODBUS request/reply PDUs. This protocol specification document describes the function codes used within the framework of MODBUS transactions." Modbus Organization, Inc. http://www.modbus.org/specs. php.

4. For a more complete assessment of the security capabilities of the IEC 61850 standard, see Smart Grid Interoperability Panel – Cyber Security Working Group Standards Review, Phase 1 Report, October 7, 2010, found at http://collaborate.nist.gov/twiki-sggrid/pub/SmartGrid/CSCTGStandards/StandardsReviewPhase-1Report. pdf.

5. One recent example of this is the Edge Control Node device being marketed by Echelon Corporation. "The Edge Control Node (ECN) 7000 Series provides an open, secure, and extensible hardware platform ruggedized and purpose-built for the Smart Grid." It provides various forms of network connectivity and a robust development environment. More information can be found at http://www.echelon.com/metering/ecn.htm.

6. Evans, James W., The Interface between Automation and the Substation, in *Electric Power Substations Engineering,* edited by John D. McDonald, CRC Press, Boca Raton, FL, 2003, pp. 6–11.

7. See The Advanced Security Acceleration Project for the Smart Grid (ASAP-SG), Security Profile for Distribution Management, pp. 61–62. Found at http://www.smartgridipedia.org/images/1/1b/DM_Security_Profile_-_v0_12_-_20100816.pdf.

8. See The Advanced Security Acceleration Project for the Smart Grid (ASAP-SG), Security Pro4file for Distribution Management, p. 19. Found at http://www.smartgridipedia.org/images/1/1b/DM_Security_Profile_-_v0_12_-_20100816.pdf.

9. See The Advanced Security Acceleration Project for the Smart Grid (ASAP-SG), Security Profile for Distribution Management, p. 21. Found at http://www.smartgridipedia.org/images/1/1b/DM_Security_Profile_-_v0_12_-_20100816.pdf.

10. See The Advanced Security Acceleration Project for the Smart Grid (ASAP-SG), Security Profile for Distribution Management, p. 22. Found at http://www.smartgridipedia.org/images/1/1b/DM_Security_Profile_-_v0_12_-_20100816.pdf.

11. For a brief description of the conference and the written testimony submitted, see http://ferc.gov/EventCalendar/EventDetails.aspx?ID=5571&CalType=%20&CalendarID=116&Date=01/31/2011&View=Listview.

12. IEC 62351, Parts 1–8, Information Security for Power System Control Operations, International Electrotechnical Commission, 2006–2008.

13. IEC 60870-6-503, Part 6-503, Telecontrol Protocols Compatible with ISO Standards and ITU-T Recommendations: TASE.2 Services and Protocol, 2nd edition, 2002–2004, p. 7.

14. Steffen Fries, Hans Joachim Hof, and Maik Seewald, Enhancing IEC 62351 to improve security for energy automation in Smart Grid environments, *2010 Fifth International Conference on Internet and Web Applications and Services, IEEE,* 2010, p. 135.

15. Steffen Fries, Hans Joachim Hof, and Maik Seewald, Enhancing IEC 62351 to improve security for energy automation in Smart Grid environments, *2010 Fifth International Conference on Internet and Web Applications and Services, IEEE,* 2010, p. 136.

16. IEC 61968, Part 1, Interface Architecture and General Requirements, International Electrotechnical Commission, 2003, p. 7.

17. See IEC 61970, Energy Management System Application Program Interface (EMS-API) – Part 301: Common Information Model (CIM) Base, International Electrotechnical Commission, 2009, p. 13.

18. See IEC 61970, Energy Management System Application Program Interface (EMS-API) – Part 301: Common Information Model (CIM) Base, International Electrotechnical Commission, 2009. Note that "[t]he principal objective of the IEC 61970 series of standards is to produce standards which facilitate the integration of EMS applications developed independently by different vendors, between entire EMS systems developed independently, or between an EMS system and other systems concerned with different aspects of power system operations, such as generation or distribution management systems (DMS). This is accomplished by defining application program interfaces to enable these applications or systems access to public data and exchange information independent of how such information is represented internally."

19. Gary A. McNaughton and Robert Saint, Comparison of the MultiSpeak® Distribution Connectivity Model and the IEC Common Information Model Network Data Set, 2008, p. 2, retrieved from http://www.multispeak.org/documents/MultiSpeak_and_CIM_ DIstributech_2008_article.pdf.

20. As of this writing, Wurldtech, an industrial control system security firm based in Vancouver, BC, Canada, has proposed IEC 62443-2-4, which provides a set of criteria for performing a security certification of industrial control system products.

CHAPTER 6

Transmission Automation
Can Utilities Work Together Securely?

6.1 INTRODUCTION

The power grid connects power through a series of hubs, commonly referred to as *substations*. The totality of this represents the transmission infrastructure that is used to transmit a series of high-voltage power lines throughout the country. The transmission infrastructure solves a key problem for delivering power, given the high state of regulation that currently exists in the industry and the location of the high concentration of people throughout the country. Cities considered major population centers are the primary benefactors of high-voltage transmission because it allows utilities to buy or deliver power from remote or other locations. Moreover, building out of the transmission infrastructure is key to the future of the Smart Grid. This is primarily because renewable generation, such as windmills and solar plants, must be built in remote locations due to the aesthetics and size of these assets. To get power from these remote sites, utilities will need to build out additional transmission infrastructure to market and deliver the power generated by these assets. Over 70 percent of the nation's energy production comes from either coal or natural gas.[1] Nuclear accounts for almost 10 percent of the country's generation capacity, and 6 percent comes from oil.[2] This means that less than 15 percent of the nation's energy is a result of renewable energy sources. To truly move to a more renewable energy economy, a great deal more renewable energy resources must be developed, which will likely require a great deal more investment in the transmission infrastructure.

6.2 TRANSMISSION INFRASTRUCTURE COSTS

The cost of building transmission is significant, given the constraints and supply costs associated with its construction. The process is similar to the manner in which a highway is built. Most transmission organizations maintain a transmission planning team. The transmission planning ultimately falls under regional transmission organizations (RTOs), as illustrated in the Federal Energy Regulatory Commission's (FERC) Order 2000.[3] However, states maintain a stake in how transmission upgrades will occur that provides them with the right to approve or reject transmission proposals. There are other factors as well, including the need to potentially condemn property in order to establish a right-of-way as well as a business case or justification for why the transmission line is necessary. Moreover, there are also environmental concerns associated with this infrastructure. Ultimately, everything comes down to cost—not just for equipment, but also to make the business justification that will illustrate a benefit to various stakeholders involved in the process. And then there is the cost of building a transmission line from a construction perspective. Considering the planning, construction, and agreements that should be in place to construct transmission, the costs of such projects are usually in the hundreds of millions of dollars.

The American Recovery and Reinvestment Act of 2008 provided for more than $6 billion in loan guarantees to two federal entities: Western Area Power Administration (WAPA) and Bonneville Power Administration (BPA).[4] These organizations are sub-agencies under the U.S. Department of Energy known as Power Marketing Administrations. WAPA created the Transmission Infrastructure Program, which maintains several goals, including

- "Construct and/or upgrade transmission lines to help deliver renewable resources to market
- Select, study and/or build projects under this authority that are in the public interest[5]
- Solicit public input in identifying potential projects
- Ensure projects do not adversely impact system reliability or operations, or other statutory obligations
- Ensure projects are economically feasible and are adequate to repay project costs
- Leverage borrowing authority by partnering with others."[6]

This program will be critical to the development of transmission from renewable energy resources, which will be an inextricable part of the Smart Grid effort. The BPA has undertaken a similar effort in the Pacific Northwest. The point is that constructing such an infrastructure requires significant investment by the federal government or the tax payer. This is meant to illustrate the scope of the problem in terms of cost. Investment in transmission infrastructure should be viewed strategically because there are large up-front costs, and the benefits will likely not be realized for years to come. However, to achieve the perceived energy independence goals along with Smart Grid, this investment must be made on a massive scale in order to achieve the goals of tomorrow.

6.3 TRANSMISSION INFRASTRUCTURE FUNCTIONALITY

Transmission lines are classified by utilities in terms of voltage. Voltage is used as the measurement because, in general, the higher the voltage, the further electricity can travel and still be useful. For this reason, transmission systems typically send high voltage across wide areas in order to get electricity from generation resources into population centers. This is contrary to distribution systems, which accomplish the same objectives on the local level at much lower voltages. This is done by first generating power from a power plant that is fed into a substation, where the voltage is changed and aggregated with other power for transmission to another substation or point where voltage again must be changed. This is typically done through the use of load tap changers and transformers. Voltage is constantly increased and decreased based on the infrastructure and need. For example, a generator may generate 100 megawatts of power that is transferred out of the facility aggregately across 115-kilovolt lines. These lines carry the power out to another substation, where power from other generating facilities may converge. The line used out of the plant depends on how much power the plant will generate. The capacity of the line must be utilized in the most efficient manner. The point is to get the 100 megawatts of power from the generator to a line that can deliver the power to a distribution substation and then down to businesses and residences. Figure 6.1 illustrates how the energy is taken from generation to transmission to distribution.

Transmission infrastructure typically operates at the high and ultra-high voltage ranges from 115 kilovolts to over 800 kilovolts.[7] The following represents what is generally accepted to be transmission system voltage levels:

- *High voltage:* 33 kilovolts to about 230 kilovolts—used for transmission of power and connection to very large consumers.
- *Extra high voltage:* over 230 kilovolts, up to about 800 kilovolts—used for long-distance, very high power transmission.
- *Ultra high voltage:* higher than 800 kilovolts—used for long-distance and ultra-high power transmission.

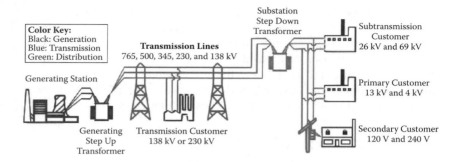

FIGURE 6.1   Generation to transmission to distribution. (From Department of Energy. https://reports.energy.gov/BlackoutFinal-Web.pdf.)

To incorporate renewable energy into the grid, the transmission system must be upgraded in order to accommodate this. This is because renewable energy is inconsistent and is probably will not be reliable. However, if the transmission system can detect and cope with these variables, it is possible to use renewable energy to provide energy to an increasing number of customers. This is seen as a direct benefit of the Smart Grid.

Windmill and solar farms are probably the most well-known sources of renewable energy. These sources of energy, however, are not aesthetically pleasing to the eye or might be considered an eyesore. As a result, there is aversion to placing these renewable energy resources in major population centers or even where they can be seen at all. The problem is that moving them into remote rural areas makes it more difficult to get those sources of power to distribution systems. To summarize, there are two basic problems with renewable energy resources such as wind and solar: they are not reliable because they depend on the weather, and people do not like the way they look so they are not placed in areas where they can directly address power needs. These problems illustrate the need to develop a transmission infrastructure that can support remote power delivery as well as the capability to address the inconsistency in power generation by renewable resources (Figures 6.2).

Creating a new transmission infrastructure is relatively simple, but extremely costly. Decisions along these lines are generally strategic in scope and are usually not profitable. For these reasons, a corporate solution is likely not viable, given the availability of the current fossil fueled generation available. To spur growth in the renewable space, the federal government and taxpayer will be required to make a significant investment in a transmission infrastructure that can support renewable sources of energy. This is not the first time that the country has faced such a challenge. The U.S. Department of Interior's Bureau of Reclamation (BOR)

(a)

(b)

FIGURE 6.2   View of potential renewable energy locations: (a) wind and (b) solar. (From NIST Government, Locke, Chu Announce Significant Steps in Smart Grid Development, http://www.nist.gov/public_affairs/techbeat/tb2009_0520.htm.)

was responsible for building out the transmission infrastructure prior to the Department of Energy Organization Act of 1977.[8] This Act not only created the Department of Energy, but also took what is commonly referred to as the power marketing administrations (PMAs) from the BOR and placed them in the Department of Energy. The power marketing administrations are responsible for marketing and delivering power from dams owned by the BOR in the western United States. There are four power marketing administrations within the Department of Energy:

1. Southeastern Power Administration—responsible for marketing electric power and energy generated at reservoirs operated by the United States Army Corps of Engineers, in the southeastern United States.[9]

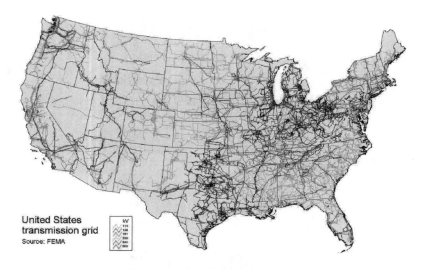

United States
transmission grid
Source: FEMA

FIGURE 6.3   United States transmission system. (Data from the Federal
Emergency Management Agency.)

2. Southwestern Power Administration—markets hydroelectric
   power in Arkansas, Kansas, Louisiana, Missouri, Oklahoma,
   and Texas from twenty-four U.S. Army Corps of Engineers mul-
   tipurpose dams.[10]
3. Western Area Power Administration—markets and delivers reliable, cost-based hydroelectric power and related services within
   a fifteen-state region of the central and western United States.[11]
4. Bonneville Power Administration—markets wholesale electrical
   power from thirty-one federal hydro projects in the Columbia
   River Basin, one nonfederal nuclear plant, and several other
   small nonfederal power plants.[12]

While the primary purpose of the PMAs is to market and deliver
power, the Bonneville (BPA) and Western Area Power Administrations
(WAPA) are participating in the American Recovery and Reinvestment
Act of 2009. Western Area Power Administration, for example, created the Transmission Infrastructure Project (TIP), whose purpose is
to "construct transmission lines to help deliver renewable resources to
market and, importantly, provides a source of funds for this activity."*
Figure 6.3 illustrates the current transmission infrastructure within
the United States. In reviewing the figure, one notices that the western

---

* Western Recovery Act Programs take from http://www.wapa.gov/recovery/about.htm.

United States does not maintain the same amount of transmission as found in Texas and the Eastern Interconnection. The WAPA and BPA work to create more transmission in order to shore up this potential gap, which will eventually be needed as more people settle in the West.

Federal programs such as TIP are designed to provide the initial infrastructure for renewable energy, which by contrast will likely enable Smart Grid technology. Between the BPA and WAPA, more than $6 billion will likely be spent over the next 20 to 30 years building out the desired infrastructure. At the same time, this infrastructure will likely be managed by cutting-edge IT-enabling assets in order to garner the most benefit from the new transmission. As time progresses, projects go forward, and new data are captured, the industry will come up with new concepts that will be designed to enable a smarter grid.

6.4 TRANSMISSION TECHNOLOGY

The Eastern Interconnect is an aggregation of many smaller transmission systems constructed under the historical vertical utility model. As such, they were intended to connect local supply with local demand. By contrast, the Western interconnect was conceived to transport electricity over long distances to bridge large centralized generation with large distributed loads.

The Eastern and Western Interconnects have limited interconnections with each other, and the Texas Interconnect is only linked with the others via direct current lines. Both the Western and Texas Interconnects are linked with Mexico, and the Eastern and Western Interconnects are strongly interconnected with Canada. All electric utilities in the mainland United States are connected with at least one other utility via these power grids.

Smart Transmission focuses on delivering the eight principle characteristics of the Modern Grid:

1. Self-healing
2. Consumer motivation
3. Attack resistant
4. Power quality
5. Distributed generation
6. Energy storage
7. Market enablement
8. Grid optimization

Smart Transmission focuses on enhancing and modernizing the transmission network capabilities to address congestion and enable wider markets and power services by enabling additional measurement and sensing, data analysis, and data control features to already complex power delivery topologies.[13] Through modeling, analysis, and real-time control capabilities, Smart Transmission will be able to address

- Congestion in the current transmission system, enabling more flexibility in delivering generation to load
- Enhanced substation automation programs
- Geographical information systems for transmission
- Wide area measurement systems (WAMSs)
- High-speed information processing
- Advanced protection and control
- Modeling, simulation, and visualization tools
- Advanced grid components for transmission
- Advanced regional operational applications

The technology needed to provide these capabilities is increasingly becoming IT dependent. For example, the control center system that monitors and manages all the remote points in a transmission balancing area is generally a major application that runs on IT to provide power management, energy management, data exchange, and automatic generation control. These systems are tied to the field assets through remote points provided by remote terminal units. In actuality, the control center system is called Supervisory Control and Data Acquisition (SCADA) and provides not only control over field assets, but is also used to monitor it. The monitoring of the system is perhaps more important than the control when it comes to smart transmission because the monitoring capability is what will notify the appropriate personnel in the event of an emergency.

The SCADA system is used in support of control center operations, primarily sustaining the energy management system (EMS), power management system (PMS), automatic generation control (AGC), and real-time data exchange (RTDE) applications. It is an isolated system protected by layers of defense, including network, process, and procedural security controls. However, all systems maintain connectivity to other systems in order to provide the appropriate level of business value. SCADA then maintains access points to various other systems and services in order to ensure that its operation supports the needs of the utility business. The needs of the utility business include AGC, SCADA development, historian, electronic tagging, RTDE, as well as control and data acquisition over substations within the balancing authority in which the SCADA system is a part. Figure 6.4 demonstrates the SCADA system and its access points, representing a traditional IT view of the system.

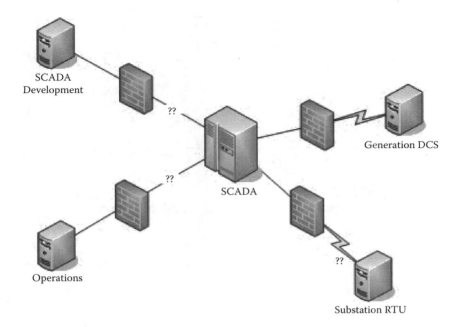

FIGURE 6.4   SCADA overview.

The SCADA system is an IT-based system used to run the EMS and
PMS. SCADA is a broad term used to define the support system for
applications that are used in a control center. The control center is the
central point within a utilities environment that is responsible for the
management of transmission and distribution operations. The EMS is
a general term associated with various advanced applications used to
provide the utility with visual insight into the status of its transmission
and distribution control areas, through data acquisition. The EMS also
maintains controlling functions that provide dispatchers within the control center with the ability to manage field assets centrally. The PMS is
similar in scope to the EMS, except that it is a suite of applications that
are used to control, monitor, and maintain power stability. Both application suites are generally used to manage the overall transmission and
distribution system from a central location.

The SCADA system is known to maintain various other systems
used to support the EMS and PMS applications. These additional systems are defined as the

- SCADA master terminal unit (MTU)
- SCADA front-end processors (FEPs)
- SCADA master database (MD)

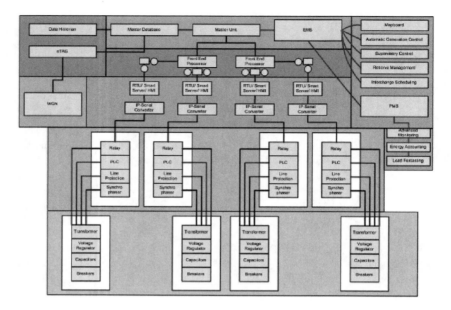

FIGURE 6.5   SCADA function overview.

6.4.1 Energy Management System

The EMS is a suite of applications used to monitor, control, and optimize the performance of the generation and transmission system. EMS applications use real-time data such as frequency, actual generation, tie-line load flows, and the plant units' controller status to provide system changes. The following applications fall under the purview of the EMS:

- Map board
- Automatic generation control
- Supervisory control
- Reserve management
- Interchange scheduling

Figure 6.5 illustrates the SCADA system and its data flows.

6.4.2 Map Board

The map board application is used to present the data aggregated by the MTU. It is essentially the dispatch board that is monitored within the control center. Whenever a disturbance or situation that needs to be responded to occurs within the generation and transmission system, the map board receives that alert-based information from the MTU and it is displayed in real-time on the map board. The map board is then a visual display of occurrences that happen within the generation, transmission, and distribution systems.

6.4.3 Automatic Generation Control (AGC)

AGC application calculates the required parameters or changes to optimize the operation of generation units. The application uses real-time data such as frequency, actual generation, tie-line load flows, and the plant unit's controller status to provide generation changes. AGC also calculates the parameters required to control the load frequency and provides the required data on demand to maintain frequency and power interchanges with neighboring systems at scheduled values.

6.4.4 Supervisory Control

Supervisory control allows dispatch to apply objectives and constraints to achieve an optimal operation of the system. In this mode, recommendations are implemented based on a predefined set of objectives. The EMS uses power flow algorithms and user-defined logics to determine the best operating settings for the system. Optimization can be used to assist energy consumers in automatically operating the system and minimizing system losses, reducing peak load consumption, or minimizing control adjustment. For energy producers, system optimization can be set to minimize generation fuel cost, optimize system operation, and maximize system security. The appropriate application of system optimization leads to a more reliable and economical operation while maintaining system voltages and equipment loading within the required range and constraints. System optimization provides intelligent load flow solutions to minimize system operating costs and maximize system performance while maximizing the value of your energy investment.

6.4.5 Contingency Reserve Management

Reliable operation of an isolated or interconnected power system requires that adequate generating capacity be available at all times while maintaining the scheduled frequency in order to avoid loss of firm loads following system contingencies. As part of the generation management and scheduling system, the reserve management application assists the system operator in continuously monitoring dynamic parameters that determine the control area's minimum reserve requirements. Reserve management maintains a constant vigil over required system reserves, including regulating reserve (spinning reserve immediately responsive to automatic generation control commands), contingency reserve (spinning and nonspinning reserves sufficient to reduce area control error to NERC performance requirements), additional reserve for interruptible imports (reserve that can be made effective), and additional reserve for on-demand obligations to other entities or control areas. Notification is issued whenever the available reserve in a class falls below the corresponding required value.

The key functions of this application include

- Identify system-wide reserve capacity requirements.
- Monitor and maintain regulating, contingency, interruptible imports, and on-demand reserves.
- Easily replace generating capacity and energy lost due to forced outages.
- Compensate for curtailment of interruptible imports from other areas to ensure reliable system operation.

6.4.6 Interchange Scheduling

Interchange scheduling (IS) provides the capability to schedule energy transfer from one control area to another while considering wheeling, scheduling ancillary services, and financial tracking of energy transactions. Dedicated for electricity power exchange and scheduling, IS incorporates energy scheduling, transaction management, and energy cost analysis and reporting. The IS provides an interface that allows for the creation of energy transactions for each location. This interface allows the scheduler to specify separate contracts for each location and assign multiple nonoverlapping schedules to each location. Interchange Area Control Error (ACE) is provided to the AGC calculation from the interchange scheduling application.

The key functions of this application include

- Energy exchange schedules
- Tariff evaluation and tracking
- Energy cost analysis

Energy exchange schedules illustrate how and when energy will be bought and sold by the utility. The utilities need this information and must coordinate it with their total generation capacity, which is measured by frequency. Utilities need to forecast how much energy they will need on a daily basis and purchase the energy amount they don't have or sell the excess energy if they desire. This obviously should be coordinated because if every utility generated energy at full capacity without consulting one another, then too much energy will flow through the system and potentially cause an overload. At the same time, too little energy will result in load shed, in that power will not potentially be available to the consumer. Tariff evaluation and tracking refer to the tariffs that must be paid as energy is bought and sold.[14] Energy cost analysis is conducted in order to determine the costs of buying and selling energy. This is an important business function as it directly correlates with the bottom line of the company. The energy cost analysis is where the business model of the company meets the operation; where operations is concerned with maintaining a reliable set of power, the business is concerned with how to generate revenue from this. The more the energy costs, the more there will be an adverse effect on the business; the less it costs, the more beneficial the impact will be, thus making this a very important application for the overall viability of the utility company.

6.4.7 SCADA Master Terminal Unit

The SCADA MTU is the primary server used to process SCADA data. It is considered the core system within the control center and generally houses the EMS and PMS applications. It is the final processor used to manage information into and out of the master database. The MTU is also the station that dispatches users to perform controlling functions on field components (breakers, switches, etc.).

6.4.8 SCADA Front-End Processor

The SCADA FEP is the secondary processor used to process SCADA data. As there is an enormous amount of data that must be processed from transmission and distribution operations, the FEP acts as a pre-processor, only providing the MTU the information that it is programmed to require and storing the rest to the master database. The FEP also acts as the terminal server used to access the HMI (human-machine interface) or RTU (remote terminal unit) within a substation. The FEP is usually in a DMZ (demilitarized zone) with inbound connectivity to the MTU. An IT-based firewall or access control list (ACL) will generally broker access to and from the MTU, in an attempt to isolate both systems. This is generally because there are a number of FEPs that provide data aggregation from various points across a number of substations.

The FEP is programmed to acquire and process data from field points and deliver to the MTU the information required by both the EMS and PMS applications. All other information is stored in the master database and eventually the data historian. In addition to this role as aggregator, the FEP provides the primary dispatch interface into the RTU or HMI within the substations under its purview.

6.5 TRANSMISSION SUBSTATIONS

Transmission substations are used to connect together two or more transmission lines. In some cases they can be used to step down voltage from one line in order to distribute it to lower voltage lines in an effort to get energy out to more remote locations. Transmission substations are the endpoints to which the control center system controls and monitors. It is the intelligent electronic devices (IEDs) within the substations that are managed by the control center system. Communications to these devices are done over operational networks. These networks can include point-to-point communications on a number of protocols or even through terminal servers. The key is that access is provided to IEDs, which are then used to control switchyard functions. For example, switching represents a key activity in the transmission substation. Switching is where the transmission lines are energized or de-energized in order to conduct maintenance. For example, it may also become necessary to de-energize a line because of a fault. The line must be de-energized so that it can be worked on or replaced. If the line were energized during this process, it could potentially result in death for the staff member working on the line. The IEDs provide the connectivity to the switchyard components, which are used to switch energy. Moreover, the IED components are effectively IT systems that are implemented to support various switching requirements within the substation.

IEDs are controlled by HMIs or man-machine interfaces (MMIs), which are meant to replace the old manual control panels (Figure 6.6). The HMIs are effectively applications running on embedded operating system platforms.

These applications are programmed to replicate what a substation operator would have to do traditionally in order to control a relay, for example. Where there was a knob on the actual panel that, when turned, controlled the relay, now that process is replicated on the HMI. The HMI is then the digitization of the substation panel. And in an effort to allow the HMI to communicate with the IEDs, TCP/IP-based networking is used to accomplish this generally in the substation only. One of the core benefits of the Smart Grid is the reduction in the amount of staff needed to support the power grid. This benefit can be directly tied to

(a)

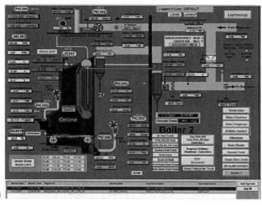

(b)

FIGURE 6.6 HMIs (a) versus traditional (b) control panels. (From (a) from OSHA Digital Bond, http://www.osha.gov/SLTC/etools/electric_power/images/control_panel2.jpg. (b) from OSHA Digital Bond http://digibond.wpengine.netdna-cdn.com/wp-content/uploads/2011/05/hmi.jpg.)

the HMIs, in that remote access is starting to be allowed to the HMI—not only from the corporate and operations network, but potentially for users working from their homes. The point is that the risk of compromise expands as the network expands.

6.5.1 Synchrophasors as IEDs

Phasor measurement units (PMUs), also known as synchrophasors, represent the industry's ability to take measurements across the power system on a consistent time basis.[15] The data provided by the synchro-phasor will likely allow the utility industry to develop additional pro-cesses and procedures that can be used to deliver many of the smart transmission benefits: self-healing, asset optimization, etc. As is the case in cyber security, the goal of Smart Transmission and the greater grid is to create intelligence that can sift through all the data related to power and create condition-based alerting that will automatically trigger con-trols to prevent disaster or optimize efficiency. A utility isolates power between two spaces today by opening a breaker. The isolation of power might be necessary because of some fault or condition that makes it required in order to correct the problem. There is a great deal of coor-dination and notification that must take place during such an event. But with information from the synchrophasor, it might be possible to pro-gram a computer to conduct the notifications, receive and acknowledge responses, and automatically open up the breaker in order to isolate the fault, which would represent the concept of self-healing.

6.5.2 Relays as IEDs

Relays are used to control switches in substation yards. They are most commonly referred to as the system that opens and closes a breaker within a substation yard. As energy is constant and ever flowing, break-ing a connection in an electrical circuit will cause electricity to stop flowing. This is similar to the concept of a breaker on the side of a house; when you move the circuit breaker, the power associated with the selected breaker is broken and the power stops flowing. A relay would be a remote device that could be used to switch the breaker. For example, in a Smart Grid, an HMI, perhaps on the thermostat, could be accessed and used to trigger the switching of the breaker through the relay.

There are a multitude of relay types that provide different func-tions, depending on the function needed. Overload protection relays, for instance, provide overcurrent protection that prevents motors from overloading. For example, if a motor in a transformer began to spin so much that the temperature was heating to a level that would cause the transformer to fail, the overload protection relay might trip it offline, effectively making the transformer stop working until the problem can be solved. You can imagine that with all the different components in a substation yard needed to control the flow of energy, many different relay types are needed to monitor the health and condition of these com-ponents. These devices start to become smart when computers are used basically to manage the relays in an automated fashion. For example, if the transformer gives an indication through the overload relay that an overload is about to occur, the computer may tell the relay to trip off.

It may at the same time have to tell another computer that it is about to do that and a transformer will be lost, so this will change the condition of the operation of the substation; a breaker may have to be tripped and the substation could be taken out of commission. There are a great many number of variables associated with such an event.

6.5.3 Programmable Logic Controllers as IEDs

Programmable logic controllers (PLCs) are used to effectively auto-mate electromechanical processes. PLCs are designed to be computers that can operate under severe conditions, such as those associated with industrial control environments. They connect to sensors that provide them with the statuses that they seek; and based on the thresholds pro-grammed into them, they may take action. For example in special pro-tection schemes, PLCs are sometimes programmed to react to anomalies or fluctuations within the power system. If a fault is detected, the PLC might be programmed to tell a relay to open a breaker. These PLCs are usually programmed by software through an HMI or MMI.

6.5.4 RTUs as IEDs

Remote terminal units (RTUs) interface with the physical components of control systems by sending and receiving telemetry data to and from connected components. These messages are basically brokered by the RTU providing an interface for monitoring and control of physical com-ponents monitored and controlled by the control system. They are most commonly associated with SCADA systems as the remote access sys-tem that provides telemetry data. Telemetry is effectively a mechanism to execute remote measurement and reporting of information that is needed in order to support the overall system. A SCADA system might have many RTUs connected to many industrial control assets, such as substations, generation facilities, etc., so that it can monitor and control some of the assets needed to support the overall system.

6.6 SMART TRANSMISSION CYBER SECURITY

The primary danger as it relates to cyber security is likely in the security of the information being used to enable smart transmission. Traditionally, utilities have been concerned about the availability of the system. The system must be available in order for it to work and provide the ben-efits to energy consumers. The following attributes of the Smart Grid illustrate functions that are needed in order to provide the resiliency that utility companies seek:

- Capability for self-healing
- Power quality
- Distributed generation
- Energy storage

Self-healing is an important aspect of the transmission system because it will ensure that when transmission is affected, that the system will automatically take corrective action. Power quality is important because the quality of power being provided through the transmission system will have a direct correlation with whether or not it can provide adequate power. If the quality of the power is high, then the transmission system will provide high-quality service; but if it is low, then problems might result. The electric grid already functions in a distributed model, where generating facilities are distributed around the country. The biggest change to this as a result of a Smart Grid is that we may start to see more local generation and more renewable generation on a larger scale. Local generation would be individual windmills on houses and in yards that put energy back into the grid. As the output of an individual generator at a home is significantly lower than a large number of windmills somewhere in the Midwest, this generation will must be compensated for. Distributed generation is thus going to become more complex, so that it can further additional reliability and resiliency. Energy storage as it becomes a reality will also contribute to the reliability of the grid because if generation is taken out, then other stored energy will be available.

While these programs will all augment the availability of the grid, they do not necessarily address the integrity of the automation that will be implemented to augment this reliability. As all of these functions will most likely result in the implementation of a software application to control these devices, what happens when the software is corrupted by a cyber security threat? As the primary mission of power grid support is safety and reliability, contingency scenarios and remediation processes have been well planned and thought out for years. But what will a utility do if a threat compromises their systems without knowing that the event occurred and then they start to invalidate the data that the utility company's automated systems depend on to provide functionality? For example, if an HMI tells a breaker to not trip when an operator or application tells it to trip based on some condition of the grid. While the operator thinks that it is tripping the breaker open, in reality nothing happens and an impact will be realized. The real threat to Smart Grid is that many of such programs could eventually make their way into the system and cause a systematic problem as a result—and on a grand scale.

6.6.1 Control Center Cyber Security

The control center system maintains the greatest amount of IT infrastructure and is generally considered the management center for each balancing authority area within each applicable utility. The control center maintains a high degree of coordination with other utility control centers as it is generally used to monitor and manage electricity. There have been many debates about the cyber security posture of the control center system, most notably the fact that it is difficult to implement various technical security controls. Because of resiliency issues, the control center systems many times cannot be scanned for vulnerabilities. This means that vulnerabilities known to exist may not even be looked at within the control center system environment. The problem of course is that not knowing what can affect the system poses a heightened risk. The compensating control that is cited to mitigate risks like these includes the fact that the control center system is isolated from the rest of the utility network.

Generally, the control center system is located at the core of any utility network where access is controlled into and out of it by a firewall. The firewall typically maintains point-to-point specific rule sets that are meant to prevent unauthorized access into and out of the system. The firewall represents the key control that secures this infrastructure. It is generally accepted that this prevents security problems within the control center system network. The control center system network within the firewall generally has few security controls applied to it because of the resiliency and reliability issues that can stem from implementing such solutions. Cyber security controls represent the threat when it comes to interior networks within the control center system, which is why the focus is on firewall protection. Using a vulnerability scanner, for example, has the tendency to bring down a control center system, or several components of it that may cause a disruption. Typically in IT environments, this is acceptable because it identifies a potential issue. In the control center system environment however, this is a rather large problem because bringing down the system, even for a few minutes, can represent a problem for the overall grid.

If a control center system goes down, the affected utility might end up not being able to see what is going on within the grid. And as monitoring for power fluctuations is a major component of the energy management system, not being able to measure frequency can be a big problem. As the main goal of the balancing authority is to keep frequency at an acceptable level, no visibility into this can result in a power disruption that may destabilize the overall grid. When you start to have computers making decisions automatically in the Smart Grid, this enhances the problem because they may not be available to make key decisions. At the same time, vulnerabilities within the control center system must be known because threats could potentially exploit those weaknesses and create the same situation. What has protected the network thus far is its segmentation from the rest of the network; however, due to cost pressures and the increasing demand for more advanced IT solutions, Smart Grid has begun decaying these barriers.

The barriers themselves are effective against traditional threats, which is what the firewall was originally meant to mitigate. However, as threats advance and networks converge, there are increasing dangers for the control center system environment. The notion of an advanced persistent threat (APT) has meant little to organizations outside of banks and the federal government. However, this particular threat represents a grave risk to control center systems because they are advanced to the point where the firewall means little in terms of blocking access. APTs use unconventional methods to exploit weaknesses in an infrastructure in order to carry out their objective. They are usually executed by nation-states and other well-funded entities that specifically target certain assets and pursue them for an extended period of time. In this case, they may want access to the control center system to effectively maintain the capability to shut it down. While there are no examples where this has actually occurred and the probability is low, the capability in itself represents the risk. And the problem is large because there really is no way to require control center systems to monitor and look for this threat because they are run, in large part, by private independent companies. And because nothing has ever happened, there is no precedence. As a result, the utilities likely view this threat as low, even though the impact can be extremely high.

North Korea, for example, might decide that it wants the ability to control South Korea's power system, and they might pay an internal employee through bribes to simply insert a USB key into their system for several minutes. Because the systems likely do not have auto run disabled and vulnerabilities are unknown, it is possible to put a virus on the USB drive that can sit dormant for weeks, months, or even years. They might even be able to accomplish this with the corporate network team that manages other systems with access to the control center system. The threat may not use the capability right away, but having the ability to access South Korea's power system can represent a real problem. This is because if North Korea ever decided to invade South Korea, they might start by using their capability to turn out the lights and then launch their preemptive strike.

While this is primarily a military issue, it does seem to bleed into the private sector in a manner never really before seen. Traditionally, government contracting companies and manufacturers are the only organizations that would have to be mobilized for a potential war. Now we are talking small municipalities who run power systems that must be hardened to prevent potential risks of cyber-attack from afar. And these organizations may not have the funding to cover such prevention, nor is the federal government able to help them financially. So we are asking our private companies to provide state-of-the-art defense for which they do not have funding.

The North American Electric Reliability Corporation (NERC) is the only entity that can really regulate the power industry for cyber security and has created the Critical Infrastructure Protection (CIP) standards to do so. The methodology developed by NERC is meant to encompass all utility control center systems whose demise would affect the overall bulk electric system. In other words, they are trying to prevent cyber-disaster impact on a national level versus the state and local levels. NERC has established six criteria that must be met in order for any utility system to be a "critical asset." When a system is declared a "critical asset," then its cyber-systems that are used to support it and are considered critical are required to be in compliance with the standards. The following represent NERC's definition of criteria that might make a system "critical."

1. If the asset has been identified as "critical" because it is a control center or backup control center, including facilities at master and remote sites that provide monitoring and control, automatic generation control, real-time power system modeling, or real-time inter-utility data exchange, then all cyber-assets that are used to support those functions are critical cyber-assets.[16]
2. If the asset has been identified as "critical" because it is a transmission substation that supports the reliable operation of the bulk electric system, then cyber-assets within the asset are evaluated in order to determine whether or not they are essential to the operation of the transmission substation in support of the reliable operation of the bulk electric system, for the reason that it was established to be critical.[17]
3. If the asset has been identified as "critical" because it is a generation resource that supports the reliable operation of the bulk electric system, then cyber-assets within the asset are evaluated in order to determine whether or not they are essential to the operation of the generation resource in support of the reliable operation of the bulk electric system, for the reason that it was established to be critical.[18]
4. If the asset has been identified as "critical" because it is critical to system restoration, including blackstart generators and substations in the electrical path of transmission lines used for initial system restoration, then cyber-assets within the asset are evaluated in order to determine whether or not they are essential to systems restoration.[19]

5. If the asset has been identified as "critical" because it is critical to automatic load shedding under a common control system capable of shedding 300 megawatts or more, then cyber-assets within the asset are evaluated in order to determine whether or not they are essential to systems restoration.[20]

6. If the asset has been identified as "critical" because it is a special protection system (SPS) (also called a remedial action scheme) that supports the reliable operation of the bulk electric system, then cyber-assets are evaluated against these criteria in order to determine whether or not they are essential to the operation of the remedial action scheme (RAS).[21] The RAS is further defined as "[a]n automatic protection system designed to detect abnormal or predetermined system conditions, and take corrective actions other than and/or in addition to the isolation of faulted components to maintain system reliability. Such action may include changes in demand, generation (MW and Mvar), or system configuration to maintain system stability, acceptable voltage, or power flows. An SPS does not include (a) under frequency or under voltage load shedding or (b) fault conditions that must be isolated (c) out-of-step relaying (not designed as an integral part of an SPS)."[22]

The control center's primary interface is in the map board display. Dispatchers depend on a map board application to show them where problems reside on the grid for which they are responsible for managing. If the map board display were to illustrate information that is inaccurate, it may be possible that this information could be taken as data that requires action. For example, if the map board shows the dispatcher that a problem exists on a line somewhere, then the dispatcher may call in resources to investigate it, when in reality nothing actually happened. This may seem like a harmless event; however, considering the resources that are required to respond to such an event, it will have a limited impact. And if a threat is able to manipulate data to create a problem like this, then the dispatchers cannot trust the information they are looking at. If dispatchers cannot trust the information they use to respond to events on the grid, then they may not be able to effectively do their job. The primary strategy that should be implemented to combat this potential problem is that cyber security professionals need to work to build a climate where data integrity protection is in place to ensure the validity of the information.

This is similar to a hashing of data approach. Information that requires a response on the part of a dispatcher should probably be validated in a manner that ensures that values have not been manipulated before they reach the control interface. For example, if a sensor sends an alarm to SCADA claiming that a problem exists, this may need to occur in a manner that ensures, most importantly, that the alarm actually came from the sensor. And the sensor should have a mechanism to ensure that it is sending the alarm based on a real condition. This should be the mindset when developing integrity controls for the power grid, that is, provide assurance that integrity will not be compromised. This is the best method to ensure reliability and, most importantly, to secure the infrastructure.

6.6.2 Transmission Substation Cyber Security

Transmission substations essentially house the assets that are used to provide functionality for the overall transmission system. While SCADA or the control center system does maintain the ability to monitor and control, it is in the substation that the assets exist that must be monitored and controlled. The primary concern within the substation is that the IEDs are not capable of adhering to sound security practices. And these systems will be controlled by HMIs and smart applications designed to carry out functions that would normally require an operator to set an application to do something. NERC CIP was likely developed to mitigate the risk of this gap. According to the methodology, if an asset is considered critical, computers controlling the assets for which it was declared to be critical will end up requiring protection. This protection is designed to be basic in scope, but can be used to interpret further security needs. For example, CIP 007.R4, Malicious Code Prevention, would basically require the implementation of an anti-virus solution to the declared "Critical Cyber Asset." While anti-virus provides a level of protection against basic threats, it does not really address advanced threats. For example, anti-virus will only detect events for which the vendor provides a signature. So new threats for which signatures do not exist will likely go undetected. Additional controls such as Security Status Monitoring in CIP 007.R6 require that the critical cyber-assets are monitored for potential security events but these are not clearly defined. The fact of the matter is that the standards should not be used to secure the environment, but rather to illustrate why a cyber security infrastructure needs to be implemented to augment security. Instead of approaching the standard from the perspective that anti-virus needs to be implemented to mitigate the risks associated with malicious code protection, we should ask ourselves, what could we implement to mitigate the risks of malicious code protection?

Looking at this from a realistic perspective, the assets will need some mechanism to monitor for network security threats, validation of access controls, anti-virus controls, code review process (vendor provided) for software installed, and system log monitoring, all correlated with one another to identify whether or not a real threat exists. Having a good cyber security manager to interpret the requirements will likely lead to a more robust cyber security posture. Depending on standards to tell you what you need to do to secure an infrastructure will likely result in the implementation of ineffective security controls. As Smart Grid is all about aggregating information and programming devices to make decisions about that information, cyber security should be based on the same premise. Threat information should be identified, analyzed, and then used to make decisions on how to secure against it.

The challenge in transmission substations specifically is to identify everything that can potentially go wrong and then attempt to determine how those things can happen. Going back to the previous example, if data can be invalidated that would result in operators initiating operations that do the opposite of what is intended, then security controls must be put in place to mitigate this risk. We may want to think about, for instance, security controls that provide data assurance that tampering has not occurred. This may be accomplished by monitoring for all changes to configurations across all assets, tying them to the change control process, and then heavily monitoring for change impact associated with each system. The configuration management process then should be dramatically improved in order to deal with advanced threats that can potentially create data integrity issues. From a reliability perspective, operators need to know for sure that they can trust the computers to execute the functions they need—while it should be cyber security's responsibility to ensure that those functions are never altered.

6.7 STRATEGIES FOR SECURING THE TRANSMISSION SYSTEM

The question then becomes: When an asset is declared "critical," do the security controls adequately protect the asset from threats, both general and advanced? Looking at the problem more deeply, if the cyber-assets that support the critical asset have no security controls, then it becomes problematic to deal with basic threats that may compromise the control center system network, for example. The control center system may maintain connectivity to corporate assets for support, such as account management, backup, and recovery, thus illustrating that there are potential vectors by which a threat on the corporate network can cross into the control center system network. The current mitigation for this risk is a focus on the air gap between the control center system and the corporate networks. The APT is much more sophisticated and deliberate, requiring much more attention for mitigation.

NERC's security controls are process related in nature and many of them can be unified together in order to illustrate larger processes. For example, the change control and configuration management requirement in CIP 003.R6 basically requires the entity to submit and approve all changes and to have a mechanism to monitor for configuration changes. What it really does not discuss is how those two key requirements tie together, and also how other requirements feed into this core requirement. If the entity was submitting and approving all changes, then all configuration changes could also be documented to supplement the change control request, meaning that every change in configuration should be tied to a change control ticket. If that were the case, then nothing on the system could change without a ticket. And if there were no ticket, then the instance must be investigated. There are a variety of tools on the market that will monitor configuration settings for a number of platforms. If those could be linked to the actual changes, then the threats could be more easily identified.

Another strategy that can be used to secure a control center system infrastructure is to actually implement, monitor, and review intrusion detection system (IDS) information. Many times, the IDS is not reviewed because of the sheer amount of data it retrieves. As a result, vendors have begun providing services where the system is tuned for your environment and even provides the ability to only allow certain types of traffic.

Managing access to the control system environment is probably the most effective way to mitigate the risks associated with the human threat. All users should have their physical and logical access to all control systems approved by someone who has been trained to give that approval. For example, the operations manager would be someone within a utility environment that would be qualified to decide whether or not someone needs physical access to a set of control system assets. This is because the operations manager understands the environment and also has a good understanding of who should need access to a certain area. Using the control center as an example, dispatchers need continuous access to the control center because it is where they are required to sit in order to do their job. However, IT administrators may not need continuous access to the area because they only enter it when the dispatchers have IT-related problems and issues. By the same token, the dispatcher probably does not need continuous access to a data center. These roles from a physical access perspective must be clearly defined so that the principle of least privilege can be used in the access management process. This would be the same for logical access controls that are likely more difficult to address because of the size and scope of the effort; there are many more computers with logical access than physical doors.

NERC CIP does a good job of addressing access management through a series of standards (CIP 004.R4, CIP 005.R2, CIP 007.R5, and CIP 006.R2). Making employees undergo background checks before they are provided access to critical areas is a good idea so that the utility company can understand the risks of allowing certain employees access to physical and logical assets. Training is another need because it forces the employee to understand the policies surrounding the assets before access is provided. Specialized training, however, should also be required to further refine this access process. For example, if an employee is being granted logical access to a set of computers that are used in automation, then those employees should be trained on what privileged access means and what types of impacts that can potentially be created as a result of doing something improper with that access type. This takes care of several problems and enables thought around cyber security. In this case, employees will or should take their provided access seriously because of the training required. This includes what they need to know as far as understanding the environment they are working in, and it will also help cyber security professionals understand what they need to defend against. If system administrators logging on to a real-time asset understand that they are not allowed to reboot the system unless it is during an outage, then there should be a cyber security control where the system cannot be shut down without special permission from the operations manager or other appropriate authority. Perhaps the system is configured to disallow system shutdown unless done by a certain user ID that is maintained by the operations manager. It is this thought process that will lead to a more secure system. Cyber security standards should be used to help in getting an infrastructure in place to enable this, but not depend on it for security.

Cyber security is a strategic endeavor, and it is important for people to understand that it will not happen overnight, but over time. Cyber security for the transmission system must evolve to support the reliability and needs of the system owners as this will provide the confidence and trust that will be necessary for the implementation of cyber security to defend against the growing threats. This will allow cyber security professionals charged with protecting the Smart Grid to evolve to a point where they understand the system well enough to know what can be done to destabilize it. This knowledge will only come from working closely with system operators.

ENDNOTES

1. For more information on grid infrastructure costs, see U.S. Energy Information Administration Independent Statistics and Analysis at http://www.eia.doe.gov/cneaf/electricity/epa/epat2p2.html.

2. For more information on grid infrastructure costs, see U.S. Energy Information Administration Independent Statistics and Analysis at http://www.eia.doe.gov/cneaf/electricity/epa/epat2p2.html.

3. United States of America Federal Energy Regulatory Commission, 18 CFR Part 35 Docket No. RM99-2-000; Order No. 2000, Regional Transmission Organizations. December 20, 1999.

4. Energy Provisions in the American Recovery and Reinvestment Act of 2009 (P.L. 111-5) March 3, 2009.

5. Western's Recovery Act Programs, retrieved from http://www.wapa.gov/recovery/about.htm.

6. Western's Recovery Act Programs, retrieved from http://www.wapa.gov/recovery/about.htm.

7. See, for example, Demetrius T. Paris and F. Kenneth Hurd, *Basic Electromagnetic Theory*, McGraw-Hill, New York, 1969.

8. Act of August 4, 1977, Pub. L. 95-91, 91 Stat. 565.

9. For more information on Southeastern Power Administration, see http://www.sepa.doe.gov/.

10. For more information on Southwester Power Administration, see http://www.swpa.gov/.

11. For more information on Western Area Power Administration, see http://www.wapa.gov.

12. For more information on Bonneville Power Administration, see http://www.bpa.gov.

13. For more information on the Modern Grid Strategy, see http://www.netl.doe.gov/smartgrid/.

14. See, for example, San Diego Gas & Electric Company Schedule Statin Power Self-Supply (SPSS), February 4, 2010, retrieved from http://www.sdge.com/tm2/pdf/ELEC_ELEC-SCHEDS_SPSS.pdf.

15. Reynaldo Francisco Nuqui, State Estimation and Voltage Security Monitoring Using Synchronized Phasor Measurements. Dissertation submitted to the Faculty of the Virginia Polytechnic Institute and State University, July 2, 2001 (Blacksburg, VA), retrieved from http://scholar.lib.vt.edu/theses/available/etd-07122001-030152/unrestricted/rnuqui_dissertation.pdf.

16. For more information on NERC CIP 002, see http://www.nerc.com/files/CIP-002-3.pdf.

17. For more information on NERC CIP 002, see http://www.nerc.com/files/CIP-002-3.pdf.

18. For more information on NERC CIP 002, see http://www.nerc.com/files/CIP-002-3.pdf.

19. For more information on NERC CIP 002, see http://www.nerc.com/files/CIP-002-3.pdf.

20. For more information on NERC CIP 002, see http://www.nerc.com/files/CIP-002-3.pdf.

21. For more information on NERC CIP 002, see http://www.nerc.com/files/CIP-002-3.pdf.
22. For more information on NERC CIP 002, see http://www.nerc.com/files/CIP-002-3.pdf.

Distributed Generation and Micro-Grids

Can Distributed Systems Work Together?

7.1 INTRODUCTION

Power generation is perhaps the backbone of the electricity industry. It is the place where energy is essentially created and introduced into the power system. Traditional generation has relied on large power plants that create energy using coal, oil, natural gas, or nuclear. This infrastructure is effectively distributed because there are power plants across the nation supplying the power system energy in a distributed manner. In fact, this distributed power system contributes to the stability of the entire system because if one plant fails, there are others in other parts of the country that can compensate for the loss of power through the transmission infrastructure. This obviously requires coordination on the part of the operators, but doing so across vast areas demonstrates the complexity of the system. The point is that the distributed nature of the power generation system contributes to the resiliency of the overall system.

Smart Grid introduces more distribution among physical assets within the power system. When you consider the fact that distributed generation in Smart Grid means that the consumer will have greater flexibility and opportunity to generate electricity and sell it back into the grid, the benefit of it can be understood. While this already occurs typically with solar panels on the roofs of consumers, the use of other renewable sources of generation are being contemplated to determine how feasible they will be in supporting the grid. While renewable energy resources represent more generation that can expand the distributed nature of the system, they can also be a major problem because they depend on the weather and cannot be used on a consistent basis. These generation resources are considered distributed energy resources (DERs), which represent many small generation assets in contrast to major generation, which represents a small number of large assets.

7.2 MAJOR GENERATION RESOURCES

There are a few forms of what is considered major generation, including nuclear, coal, natural gas, and hydro generation plants. The compilation of coal and natural gas is considered fossil generation, meaning that fossil fuels are used to produce energy while hydro generation uses water to produce energy. And finally, nuclear is yet another category because of its use of nuclear technology to produce energy. These plants are considered major because they produce the largest amount of energy. The Palo Verde Nuclear generating station outside of Phoenix, Arizona, for example, generates nearly 4,000 megawatts of energy, serving more than four million people.[1] Comparatively, the largest coal plant in the United States is Plant Scherer, north of Macon, Georgia, and generates almost 3,500 megawatts of energy.[2] The largest hydro plant in the United States, Grand Coulee, can produce up to 6,800 megawatts of electricity.[3] It is worth mentioning that the largest hydroelectric dam in the world, Three Gorges Dam in China, can produce upwards of 22,000 megawatts.[4]

Compared to DERs, this is massive, and it is easy to imagine how important major generation is to the public. If the entire system were made up of small DERs, it would take an extremely large number to offset the need for major energy. Major generation is also distributed across a vast plane of resources throughout the country. They are found in many rural areas, but also near major population centers. A hypothetical example to provide the context of this could be that if roughly 1,000 megawatts is equivalent to serving load to 1,000,000 customers, then it takes upwards of 150,000 megawatts to sustain the load to the general population of the United States, assuming there are approximately 150 million separate housing units and discrete business facilities in the United States.

7.3 MAJOR GENERATION COSTS

The cost of major generation is significant. The general cost to build a pair of nuclear reactors is in the range of more than $6 billion, and this does not include the amount of time it would take to complete such a project, nor does it incorporate inflation.[5] Building out a nuclear power plant also requires federal government intervention because of the potential harm to public safety in regard to the possibility of a radiation leak. Nuclear power plants are, in effect, not insurable without appropriate federal intervention, which is mostly because of potential impacts. In fact, the Price-Anderson Act of 1957 placed a cap on the amount that could be claimed during a nuclear power plant disaster.[6] As such, nuclear power plants are subjected to very strict safety guidelines set by the Nuclear Regulatory Commission. The costs for coal and natural gas plants are not nearly as significant, but economic environmental regulations are driving up the costs. And finally, hydro projects cannot even really be done on a massive scale in the United States for several reasons, one of which includes environmental concerns. Moreover, there are growing lake level problems within the United States that have a direct impact on the ability to generate adequate electricity from hydro.

Compliance issues also represent a challenge to the cost of managing coal, natural gas, and hydro generation capabilities. Compliance is an issue not only from a reliability perspective, but also from an environmental perspective. Coal and natural gas plants, for example, are heavily regulated by the U.S. Environmental Protection Agency (EPA). Recently, many coal plants have introduced the concept of "scrubbers" that are designed to control the pollution emitted.[7] Growing compliance concerns are increasing the cost of generation from fossil fuels, and it may be likely in the future that these costs could grow to the point where it becomes more economically feasible to use renewable sources of energy.

Table 7.1 illustrates the cost of major generation alongside several forms of renewable energy. The table is meant to illustrate the difference in cost between sources of energy on a per-kilowatt basis. Because 1 megawatt-hour (MWh) is equal to 1,000 kilowatt-hours (kWh), the cost of each energy resource can be easily calculated. If the cost of natural gas, for example, is 3.9 cents per kilowatt-hour, then the cost of 1 megawatt-hour is $39. So in a plant that generates 1,000 megawatt-hours of

TABLE 7.1   Generation Cost Matrix[a]

Generation Source	Cost per kWh
Gas	3.9–4.4 cents
Coal	4.8–5.5 cents
Nuclear	11.1–14.5 cents
Wind	4.0–6.0 cents
Hydro	5.1–11.3 cents
Solar	15–30 cents

[a] For more information, see http://peswiki.com/index.php/Directory:Cents_Per_Kilowatt-Hour.

electricity, the total expenditure on energy output is roughly $39,000 per hour. From here, the cost becomes a variable because there are multiple factors involved in the calculation of costs, such as what the base load of the plant is.

From Table 7.1 it is easy to see that solar power is the most expensive power, but nuclear represents the largest cost from a major power generation perspective. And the variables come into play with renewable energy because their ability to generate power is relatively inconsistent. For example, if the sun does not shine, then solar power cannot be created. Similarly, if the wind does not blow, then wind turbines will not spin and energy will not be created. Whereas in the case of gas, coal, and nuclear, they depend on fuel that is provided steadily. This is the key problem that Smart Grid serves to solve. A computer must be programmed to understand these inconsistencies in renewable forms of energy and should be able to make decisions based on the amount of load coming out of areas where renewable energy is located. For example, if solar is used to generate energy during the day when the sun is out, then something needs to compensate for the lack of solar energy at night. Energy storage, such as large batteries, is being explored as one possibility to store energy from unpredictable sources such as solar and wind in order to make it more predictable. However, the value of Smart Grid technology is to balance a wide variety of these energy resources to leverage them when they are available rather than try to depend on a single generation source to provide power to a particular load. This requires not only sophisticated software to help decide what resources to draw from when, but it also requires a transmission and distribution system that is able to delivery energy from multiple sources.

A hybrid solution of both renewable and traditional generation will be necessary to bridge power generation in transition to a total renewable energy source. As predictive models improve and Smart Grid becomes a reality, this can be possible. But the scope of the effort must be taken into account. As stated in previous chapters of this book, a limited number of renewable energy resources are used to supply power, and in this chapter we see that it is very complicated to bring renewable sources online because of the variables in the supply of fuel (e.g., wind and solar). Wind and solar then need to be seen as the fuel source and a supply chain infrastructure needs to be established for it. However, this fuel supply is difficult to measure because of the variables that need to be known: when the wind will blow and when the sun will shine. So weather will play a significant role in the establishment of renewable energy resources as the dominant energy resource.

Another solution would be in the improvement of energy storage. As energy is constantly flowing, it is difficult to store. Currently there is a limited amount of energy that can be stored because of the amount of capacity that would be necessary to support the vast amount of energy needed to support the power infrastructure. As it currently stands, the technology does not really exist to support what would be necessary to store energy on a massive scale. There is no storage device, for instance, that can take 1,000 megawatts of power from various wind farms and store it while the wind is blowing so that it can be redistributed when the wind is not blowing. There are various companies and research firms working to solve this problem and the breakthrough in such technology will go a long way in solving this problem. If it were possible to store these massive amounts of energy, then it becomes much easier to offset the supply of wind and solar power with the use of energy storage. One could then say that if we solve the energy storage problem, then we will be able to take full advantage of renewable energy resources.

7.3.1 Nuclear Power

Nuclear power is one the current energy resources that must be balanced. Figure 7.1 illustrates the anatomy of a nuclear power plant. The control rods are inserted into the water within the reactor. Because of the massive heat produced by the control rods, the water boils and produces what is known as concentrated steam.[8] The steam is then passed through a steam generator, which effectively passes the steam onto a turbine. The turbine spins as a result of the steam that causes the generator to run, creating electricity. The trademark of the nuclear power plant, the cooling tower, is simply used to cool the condensed water to an acceptable level that is then pumped back into the system. Each process within this system is controlled by a distributed control system (DCS).

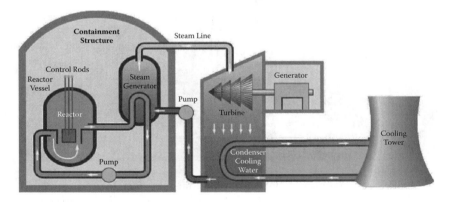

FIGURE 7.1   Nuclear power plant. (From http://www.nrc.gov/reading-rm/doc-collections/fact-sheets/3mile-isle.html#tmiview.)

The DCS is distributed, in that it is located throughout the plant and used to control each process needed to manage the production from the plant. Inserting control rods into the reactor, for example, is executed by the DCS. This process is then controlled by the DCS, which is why the DCS is also known as a process control system. The whole point of the DCS is to manage all the processes that are involved in generating electricity.

7.3.2 Coal Power

Similar to the nuclear power plant, a coal plant also produces energy by boiling water and producing steam to spin turbines. The key difference is that the coal plant produces heat to boil water by burning the coal that is supplied to the plant. Coal plants also depend on a steady supply of coal that must be mined from elsewhere, usually loaded onto trains and then shipped directly to the plants. If the coal supply were ever disrupted for an extended period of time, then the plant would eventually stop functioning. This is counter to the function of a nuclear plant, which depends on fuel rods that run constantly for years. Figure 7.2 illustrates how a coal plant functions.

Coal shipped to the plant is transported to a pulverizer, which grinds the coal into matter that can be easily burned. The ground-up

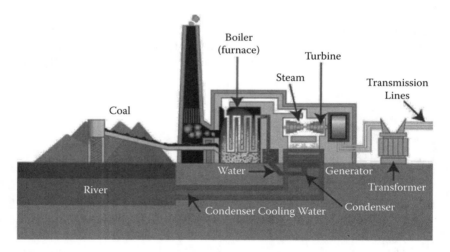

FIGURE 7.2   Coal power plant. (From a diagram of coal-produced electricity by the U.S. Department of Energy. Graphic online, courtesy of Dept of Energy. http://www.nrc.gov/reading-rm/doc-collections/fact-sheets/3mile-isle.html#tmiview.)

coal is then released into the furnace. The furnace burns the coal, producing steam, which is then taken into the turbine. Spinning of the turbine effectively causes the generator to create electricity, which is then transported out to the substation. The coal plant is also managed by the DCS, or process control system. Each process within the process of generation is effectively controlled by a computer system, which is the DCS.

7.3.3 Gas Power

The natural gas plant functions somewhat differently from both the nuclear and coal power plants. A natural gas line is brought into the plant and fed directly into what is known as combustion chambers. The combustion chamber works similar to the way an automobile functions, where the gas is burned, producing energy that spins the turbines and generates electricity. Figure 7.3 illustrates this process.

Each process is again managed by a DCS system, which ensures that everything is functioning properly. It is easy to imagine the need to monitor each process for threshold levels to ensure that none are operating above or below the intended limits. For example, to account for a situation where the gas line is blocked, there should probably be a sensor in place to tell the system what has occurred so that it can alert personnel as to what is going on.

7.3.4 Hydroelectric Generation

Hydroelectric generation is a much simpler process, in that it simply uses gravity and water to spin the turbine in order to generate electricity. Dams are generally built to stop flooding, but as a side benefit they can

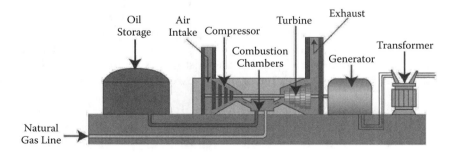

FIGURE 7.3 Natural gas power. (From Tennessee Valley Authority (Federal Government) http://www.tva.gov/power/fossil.htm.)

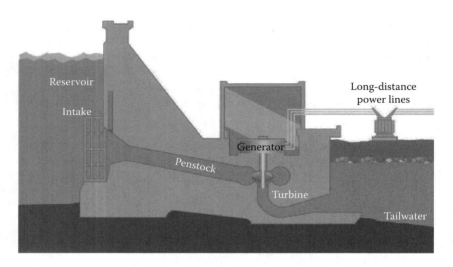

FIGURE 7.4   Hydro generation power.

be used to produce power. Because they are used to hold back water, they are generally good places to build generation capabilities. That is because the water can be allowed into a chamber that sits atop the dam. At the bottom of the dam, where the chamber ends, would be where the turbine resides. As water flows down, it spins the turbine, creating electricity. Figure 7.4 illustrates this.

The larger the dam, the more energy that can be generated. This is because the water will flow faster, causing the turbine to spin faster, ultimately generating more electricity. After the water flows through the turbine, it is sent into another outlet that allows the water to flow downstream. Another benefit of hydroelectric plants is that energy is not needed to create the energy. The gates that control access to the penstock can simply be opened manually to allow the flow of water spinning the turbine. This is why hydroelectric generation is used widely as blackstart generation. "Blackstart generation" is a term used to describe how energy would be created if there were no power to the system. Blackstart generators can usually run apart from the general power system. Again, one does not need electricity to open a penstock, as it can be done manually. Once power is started, then there is power to start other generation resources, and so on.

7.3.5 Distributed Energy Resources (DERs)

DERs effectively allow the consumer active participation in the generation of electricity. DERs are small, modular, energy generation and storage technologies that provide electric capacity or energy where you need it.[9] DERs usually produce less than 10 megawatts of energy and are connected to the local power distribution system. These systems are typically not used on the transmission system as they are not high-voltage power generation systems; they are in the very low voltage range. DERs include two types of energy:

1. Energy generation
2. Energy storage

7.4 DISTRIBUTED ENERGY RESOURCE COSTS

The cost of distributed energy resources is largely localized. This means that consumers or communities will be responsible for establishing these systems on large or small scales, or a large number of small generation systems. This is in contrast to the current system that is represented by a small number of large generation resources. DERs are comprised of energy generation systems that are meant to produce power below 10 megawatts. And energy storage technology is available to store energy on a small scale. The cost, however, is far more fragmented than with major generation. Each individual entity needing power would have to implement its own energy generation system and use it to sustain its own infrastructure. The cost of this is likely not significant enough to get them away from using major generation or energy sold to them by a utility company. Again, a hybrid solution might be the answer to this, wherein the primary source is in the DERs implemented for entities, but with reliance on major energy resources for secondary power—that is, for that energy which cannot be sustained, major generation will compensate for that lack of energy.

7.4.1 Energy Generation Systems

Energy generation refers to generation resources that can be used to create power. These are generally lower-cost alternatives to major power generation, meaning that they generate less than 10 megawatts of power. The following describes some of the different types of energy generation resources that one would see associated with a DER:

- *Wind turbines:* Wind turbines are a renewable energy resource that uses propeller-like blades that turn to produce energy. The propeller-like blade represents the turbine that, when spinning, causes the generator to produce energy. Most wind turbines are asynchronous, meaning they turn at variable speeds. Wind turbines range in electrical output from a few watts to more than 1 megawatt. Applications include remote power systems, small-scale or residential electricity production, and utility-scale power generation. DER-scale systems can cost anywhere from $1,000 per kilowatt to $3,000 per kilowatt, depending on how they are used.

- *Diesel engines:* Diesel engine generator sets (gensets) are a proven, cost-effective, extremely reliable, and widely used technology. They are manufactured in a wide range of sizes, from about 1 kilowatt up to about 10 kilowatts. They can be cycled frequently to operate as peak-load power plants or as load-following plants; they can also run in base load mode in off-grid systems. Diesel engine gensets can start at a cost of $800 per kilowatt-hour.[11]

- *Natural gas engines:* Natural gas engine gensets consist of a reciprocating (piston-driven) natural gas-fueled engine using a spark-ignition system (Otto fuel cycle) coupled to an electric generator.[12] In most other respects, natural gas engines perform similarly to diesel and dual-fuel engines, but have the potential for the lowest emissions of all types of reciprocating engines.[13] They are available in sizes from a few kilowatts to about 5 megawatts, and they cost about the same as diesel and dual-fuel engines.

- *Dual-fuel engines:* Dual-fuel engine gensets consist of a diesel-cycle engine modified to use a mixture of natural gas and diesel fuel (typically, 5 to 10 percent diesel by volume) connected to an electric generator.[14] The small amount of diesel fuel allows the use of compression ignition, and the high percentage of natural gas in the mix results in much lower emissions (and somewhat lower power output) than those of a diesel engine. In most other cost and operational respects, dual-fuel engines are comparable to diesels; they are available in sizes from a few kilowatts to about 10 megawatts at a cost similar to that of natural gas and diesel engines.

- *Combustion turbines:* Combustion turbines (also called gas turbines) burn gas or liquid fuel; hot gases expand against the blades of a rotating shaft, producing a high-speed rotary motion that drives an electric generator.[15] While they may take a few more minutes to get up to speed in comparison to reciprocating engines, gas turbines are well suited for peaking and load-following applications and for base load operation in larger sizes. The costs are somewhat higher than those of reciprocating engines, and maintenance costs are slightly lower. Turbines are efficient and relatively clean. They can easily be fitted with pollution controls to run even cleaner. The cost of this technology is slightly higher than dual-fuel, natural gas, and diesel engines.

- *Micro-turbine systems:* Micro-turbines are smaller, somewhat less efficient versions of combustion turbines, in the range of about 30 to 250 kilowatts.[16] They run on natural gas at high speeds. The electrical output of the generator is typically passed through an inverter (an electronics-based power converter, also called a power conditioning unit or PCU) to provide 60 hertz AC power. Micro-turbines are designed to be compact, affordable, reliable, modular, and simple to install. The cost of micro-turbines runs around $1,000 per kilowatt, somewhat higher than combustion turbines.

- *Fuel cells:* Fuel cells produce DC electricity by a thermochemical process in which hydrogen (H_2) is passed over an anode and air over a cathode in an electrolyte bath.[17] Fuel cells are efficient, quiet, and modular. They are available in sizes ranging from a few watts to 200 kilowatts. However, they are also considered one of the most expensive DER technologies, starting at about $5,500 per kilowatt.[18]

- *Photovoltaics:* Photovoltaic (PV) cells are thin layers of a semiconductor (usually crystalline silicon) that convert sunlight directly to DC electricity; an inverter converts the DC to standard AC power for connection to utility systems.[19] These "solar cells" are built up into panels with power ratings ranging from a few watts to about 100 watts. The panels are modular and can be configured into larger arrays to match almost any load requirement. Noise and emissions are nonexistent, and maintenance is minimal because there are no moving parts. Depending on the application, PV systems can range from $8,000 per kilowatt to $13,000 per kilowatt. Grid-connected systems typically fall in the low end of the range, while systems with battery storage constitute the high end.

7.4.2 Energy Storage Systems

Energy storage systems are used to store energy, similar to the way a rechargeable battery works. There are various forms of energy storage devices, but the technology is limited in the amount of energy that can actually be stored. This is one of the reasons why there are limitations on the distance an electric vehicle can travel, as the battery will only hold enough energy to allow a limited distance to be traveled. This is also true in energy storage systems within DERs. The following illustrate some of the energy storage devices used in DERs:

- *Uninterruptible power supply (UPS) systems:* An uninterruptible power supply (UPS) is an electrical apparatus that provides emergency power to a load when the input power source fails.[20] A UPS differs from an auxiliary or emergency power system or standby generator in that it will provide instantaneous or near-instantaneous protection from input power interruptions by means of one or more attached batteries and associated electronic circuitry for low power users, and/or by means of diesel generators and flywheels for high power users. These systems are used usually on a temporary basis to supply power instantly when it is lost.
- *Superconducting magnetic energy storage (SMES):* SMES uses a magnetic coil cooled to very low temperatures to store electric energy with little loss; like other DC devices, it uses an inverter to convert DC to AC that can be dispatched to a utility grid.[21] The cost of energy storage systems can range from $1,100 per kilowatt to $1,300 per kilowatt.
- *Flywheel systems:* Flywheels convert electric or mechanical energy into rotational energy and invert it for use when needed.[22] These are sometimes used as an alternative to or as a complement to UPS technology for large data centers and other uses. They tend to be more environmentally friendly than battery technology and often have longer usable life spans.
- *Hybrid systems:* Hybrid systems are combinations of these technologies, designed for specific or unusual applications.[23] Renewable energy technologies such as wind and solar systems, for example, depend on energy sources that cannot be dispatched. For this reason, it may be necessary to combine them in a hybrid system, such as a photovoltaic system with battery backup to collect energy for use when a facility needs it. Nonrenewable hybrid DER systems are also used; one example is a battery system packaged with a micro-turbine to ride through short outages with the batteries and use backup power from the micro-turbine for sustained outages.[24]
- *Large battery systems:* Large battery systems refer to advanced energy storage systems that have a capacity far greater than a typical UPS system. These systems typically utilize electrochemical energy storage methods, which maintain high-energy density, are flexibility, and scalable. These systems are those that will one day be used to store energy and then use it in grid

operations. This is contrary to UPS technologies, which are used primarily for emergency power. Large battery systems will be used to sustain power over an extended duration of time.

7.4.3 DER Programs

DER programs represent the goals associated with distributed energy resources. These goals are effective solutions meant to address key problems associated with DER. The following represent the list of programs associated with DER, which represent challenges that are meant to be addressed:[25]

- *Peak shaving:* Utility companies charge a premium for the use of energy at peak time periods. Peak time periods are times when most entities really need energy, which puts a strain on demand. This demand then causes the price to increase because the supply of energy becomes limited during these time periods. Take, for example, Florida, where the summers are extremely hot and air conditioning becomes a necessity during the day. The hottest point in the day may be between 2 p.m. and 4 p.m., and as a result a utility might see this as peak usage because most of the air conditioners in a given service area are in use. As such, the utility might charge a premium for using energy between those times that could be four or five times the normal rate. This provides the consumer with an incentive to not use power during those time periods. However, because of the heat, it may be necessary. This incentive calls for consumers to use a DER during these time periods to reduce their cost. This would go hand and hand with a utility company's time-of-use programs discussed in the Chapter 3. The concept of peak shaving is then that a DER would be used during peak hours to offset the cost of using utility-provided power during these times in order to save money.
- *Improved power quality:* Where power quality issues are observed, a consumer might implement a DER to offset the effect of this. Power quality would have to be improved where frequent momentary outages cannot be tolerated. Consumers might implement a UPS to mitigate the effects of this if they have a mission-critical asset, where power disruptions would cause a great deal of harm. A good example might be in a data center or hospital, where these issues could cause great harm. Power quality is important in these environments because the systems that rely on them are critical to something significant to the consumer.

An adverse effect on power quality can create a destabilizing situation for the power grid. As a result, utilizing DERs to achieve power quality can go a long way in sustaining the reliability of Smart Grid. The risk of cyber security threats that threaten the stability of major generation can be offset by the implementation of DERs. This is because DERs serve a local function that enhances power quality locally, instead of depending on power generation remotely.

- *Green power:* Green power refers to the environmental push to establish renewable energy resources as the dominant form of power generation. This push is being established because of the perception that traditional forms of power generation have an adverse impact on the environment. Consumers would in such cases implement DERs that are environmentally friendly, such as wind turbines or solar panels. This may also help in reducing the cost incurred through emissions standards, as these renewable sources of energy are not significantly affected by emissions standards.

7.5 DER CYBER SECURITY

The effect of DERs in relation to cyber security has not yet been seen or measured. These systems currently are more localized to organizations that utilize them to support their electrical needs. As these systems become integrated in the overall power infrastructure and become useful distributed resources that augment the supply of major generation, they will be connected by an ever-growing communications network. Therefore, the communications infrastructure and exposure of the networks represent the greatest risk when addressing DERs. This is mostly because they will have to interact with major generation resources and other power system assets, through the communications infrastructure. Currently, the control center only monitors major generation resources, but as Smart Grid becomes a reality, it may be necessary for them to monitor DERs or to get some input as to which DERs are available, in the event that the demand drives the necessity to create a response using the DERs. The fact that the DERs are used is not the problem, but, rather, it is really the communications with them.

Communications with DERs through the control center might become a problem because of their distributed and unsecured nature. Typically, anyone will be able to implement a DER and sell power from it back into the grid. This denotes that the utility will not control the endpoint that is used to aggregate the data from the DER. If a threat were to compromise that information from the DER supplier, how will the utility know for sure that the information they are receiving is correct? Furthermore, what would be the impact of receiving inaccurate information on power availability that is needed in a demand response situation? Questions like these must to be answered as generation resources for the grid become more and more distributed.

7.6 MICRO-GRIDS

In previous chapters we saw a description of automated systems to control and monitor loads via home area networks and the introduction of distributed energy resources located near the loads. In this chapter we put it all together by introducing the concept of a micro-grid. "Microgrids are small-scale, LV CHP [low-voltage combined heat and power] supply networks designed to supply electrical and heat loads for a small community, such as a housing estate or a suburban locality, or an academic or public community such as a university or school, a commercial area, an industrial site, a trading estate, or a municipal region. Micro-grid is essentially an active distribution network because it is the conglomerate of DG [distributed generation] systems and different loads at distribution voltage level."[26] One can look at integrated building management systems on college campuses or military bases and conclude that they are micro-grids. While they do resemble distribution systems and may leverage diesel generators for backup purposes, they tend to lack a robust generation source that can take over for the utility at a campus level when needed using a technique called "islanding," where the enclave is disconnected from the utility's grid for a period of time. This capability may also be called co-generation but is distinct from the distributed generation discussed in the prior chapter in that the generation is combined with a fairly substantial array of distribution components, such as transformers, capacitors, automated switching, and other features normally associated with a utility's distribution system. In some senses, the micro-grid is a reversion back to the days where everything was local and there was no bulk electric system. The difference, of course, is that the bulk electric system is still very much a part of the micro-grid for both cost and reliability reasons. While it has a more robust generation source, sometimes called a "microsource," a micro-grid is only intended to operate in "islanding" mode for short periods of time, such as during peak demand periods or during an emergency. The microsources are also likely to be renewable resources or more efficient fossil fuels like natural gas. One other feature of micro-grids is the ability to use the excess heat, usually in the form of steam, resulting from the generation process to provide heating for the local campus building. This was a familiar feature of early electric grids before zoning and environmental

FIGURE 7.5   Capitol Power Plant.

regulations made generation an activity that occurred far away from the load source. For example, the U.S. Capitol Power Plant in Washington, DC, used to supply electricity and steam to nearby congressional office buildings. "Instead, the plant generates steam and chilled water to heat and cool the Capitol, the Supreme Court, the Library of Congress and 19 other structures. Steam and chilled water are carried in pipes through a web of tunnels stretching from south of the Capitol to Union Station."[27] While hardly the epitome of a modern micro-grid, the Capitol Power Plant shows that the concept is hardly new (Figure 7.5).

7.6.1 Micro-Grid Functions and Smart Grid Interaction

In essence, a micro-grid should be a miniature version of an electrical grid with the exception of no transmission network because the generation is local. Otherwise, the micro-grid will have a generation source, substations for stepping down the voltage, capacitor banks for reactive power and balancing, metering at the building and possibly the load levels, building management and/or home area networks for load control, a communications network, distribution SCADA systems, demand response mechanisms, and generation planning capabilities. Many of these mechanisms would be centrally located with more modern automation. However, the functions would be similar. Finally, the micro-grid must be able to operate in synch with the larger grid. And given its size in some cases, it must carefully coordinate its activities to avoid cascading effects of any anomalies on the micro-grid from making it onto the bulk electric grid. The warnings about how smaller utilities could affect the bulk electric grid under the right circumstances apply here as well.

Because modern micro-grids are relatively untested, the precise functions are not well defined. However, at minimum, a micro-grid should be capable of acting like an electrical grid, which means that it should be able to balance generation with the load demanding energy. And because this could be potentially be done "off grid," the micro-grid will essentially have to act as its own balancing authority even if most of these functions will likely be more automated than the balancing processes performed by regional transmission organizations (RTOs) or independent system operators (ISOs). Additionally, micro-grids may need to coordinate the migration to and from the grid with the local utility if there is only a single power supplier, or possibly with an RTO or ISO if multiple utilities are involved. Depending on its size and load variance, a micro-grid should practice energy market forecasting to ensure that it is aware of the anticipated loads depending on variables like the weather and usage patterns for energy-hungry machines. Depending on the loads involved, this is just as critical when the micro-grid operator decides to operate in co-generation mode, where part of the required energy may be generated by the micro-grid and part by the utility. "Supervisory control and data acquisition (SCADA) based metering, control and protection function should be incorporated in the Microgrid CCs [central controller] and MCs [microsource controller]. Provisions must be made for system diagnostics through state estimation functions."[28]

7.6.2 Cyber Security for Micro-Grids

Notwithstanding the common functions described, micro-grids are likely to be as different as the public that will use them. For some affluent communities, a micro-grid may be a series of solar- or wind-based energy resources that provide power during sunny and windy days and rely on the power from utilities the remainder of the time. For military bases, it may mean the ability to remain functioning in the face of some of the worst natural disasters or enemy attacks. For others, it might offer greater reliability for remote communities at the end of long and unreliable distribution lines. And for some Third-World villages, it might offer electricity to residents who previously had to go without. In each of these cases, the tolerance for cyber security risks will be different. A community accustomed to electricity outages is probably more prepared to handle the effects of a cyber security attack. Alternatively, an affluent community wanting to be more environmentally friendly may be ill-equipped to handle a disruption of even a few hours. Consequently, the design of a micro-grid and the accompanying power from the utility will need to consider the various risk tolerances and the resources available. The affluent community using wind and solar will need an electricity supply from the utility that provides for all its needs and assumes that renewable resources will be unreliable. Such a model is not much different than what most communities are accustomed to. In the event of a utility outage, the community may have to go without briefly, but such occurrences are rare and can be mitigated somewhat by the renewable generation the community has deployed. On the other hand, a military base is seeking to get to as close to zero outages as possible. This is more critical as military bases may be needed more in times of disaster or an enemy threat that disables the primary power grid. The base's micro-grid should incorporate advanced load control features to ensure that the critical base functions always receive power. The communications links and electric cables should all be hardened to limit the effects of a physical attack. Additionally, the bases should avoid the use of wireless signals for critical control traffic to avoid radio frequency interference. Communication traffic should be encrypted to avoid tipping off eaves-droppers as to when more power is being used on base, possibly to pre-pare for a classified operation. Finally, the various technologies meant to secure the distribution, AMI, and HAN parts of the grid discussed in Chapters 3, 4, and 5 should be used. With respect to that remote Third-World village, the biggest risk may be energy theft unless the electricity is offered free of charge to all villagers.

While disruptions to a micro-grid are a concern, the utility's greater concern is non-utility personnel who may not be subject to the same regulation operating an electric grid that it may need to interconnect with. This presents a number of issues. Central among them are the methods that utilities and micro-grid operators will use to communicate status and control information. A utility and micro-grid may agree to a particular set of circumstances where islanding and grid connected mode is desired. Moreover, automated switches, such as reclosers, may be positioned at the interconnection to respond to disruptions on either end or to receive instructions to disconnect from either end. Dropping or adding a sizable load such as a large micro-grid requires some planning and coordination. However, many military bases and college campuses already run a mini-distribution system with a single or dual electric feed from the utility with the campus or base responsible for downstream metering of buildings, SCADA operations, and other protection measures. If a tree fell on the interconnection point, the effects would be the same as an attacker breaking into the micro-grid's communication network and initiating a disconnection from the utility feed. In some cases, the ability to connect a campus or base from the larger grid may already be a capability in place. The higher visibility of micro-grids may make them a greater target in the future, but as currently proposed, the impacts do not appear to be much greater.

7.6.3 Future of Micro-Grids

The future of micro-grids is bright as they offer greater resilience for underserved areas and even more reliability for those communities that cannot afford to be without power. Because a micro-grid is really just a smaller version of the electrical grid, the same cyber security challenges are present to some degree. The hope is that they are more manageable if contained within a more limited geography using more advanced technology to monitor status and act on problems. The danger is that the more advanced technology can be leveraged to more rapidly solve problems as well as create them. Applying a solid cyber security regimen to these issues is one way to mitigate the harm.

7.7 DISTRIBUTED CONTROL SYSTEM

The distributed control system (DCS) is an IT system used to control, monitor, and operate the power plant. It is called a distributed control system because it controls the distributed resources within the plant. For example, the boiler used to boil water in a coal plant is usually run by a boiler management system. The boiler management system application is a part of the DCS that generally has control over this area of the system. The DCS may also be used to manage the balance of plant application that coordinates all processes within the plant. The balance of plant applications basically monitors all processors to ensure that the plant is functioning properly. For example, if the process that controls the pulverizer stops functioning, the balance of the plant would notify the control operators of such an event. There are a variety of applications associated with a power plant included in the DCS. Each process within the plant will likely be associated with some application that makes up the DCS. The applications in use within the generating facility (at a coal plant, for example) might be used to control and monitor the following functions:

- Coal supply
- Boiler
- Turbine generator
- Condensers and the cooling water system
- Water treatment plant
- Ash handling systems, including precipitators
- Continuous emission monitoring

For every function, there is likely an application or program that has been made in order to support or automate that function.

The DCS network represents the network that supports the applications used to control and monitor processes within the plant. The DCS should be thought of as the IT (information technology) backbone for the plant that allows processes to function and be monitored from a central location; an example would be in the conveyor belt, where coal is dropped off for transport into the pulverizer at a coal plant. A process is in place to convey the coal from the coal supply into the pulverizer. If that conveyor belt were to stop running, then the supply of coal to the pulverizer would eventually run out and the production of energy would eventually stop at the coal plant. The DCS maintains a sensor to monitor to ensure that the conveyor belt is functioning properly and a programmable logic controller may be used to control the conveyor belt. This would allow the operator to start or stop the conveyor belt. If there were a problem, for instance, with the pulverizer, the reaction might be to stop the coal supply from sending coal to it until the problem is resolved. This illustration is meant to provide an example of what the DCS system does. The concept can be applied universally to the other processes within the plant.

7.8 SMART GRID AND DISTRIBUTED GENERATION

Distributed generation is really nothing new to the utility industry. What makes it significant in Smart Grid is that the scope will be expanded from major generation to DERs, meaning that DERs could play a major role in generation for the Smart Grid. The Smart Grid may rely on the use of the DERs to offset the lack of energy that is provided by major generation. For example, a major generation plant may generate 2,000 megawatts of energy but may need to provide 2,500 megawatts of energy. Currently, a utility may be required to buy expensive energy on the open market in order to compensate for the lack of 500 megawatts needed to support the system. With the integration of DERs into the total generating capacity of the balancing area, a utility might prearrange the purchase of energy from their consumers who do not use all of the energy that they create themselves. While a large DER may only provide, at most, 10 megawatts of energy, thousands of DERs selling small amounts of energy into the grid could grow to exceed well over the lack of energy that a utility might need.

The problem with this distributed model is the complications associated with ensuring that the consumers will be able to sell the proper amount of energy to the utility company to offset for not having the correct amount. Furthermore, there will likely be inconsistencies in the amount of energy that is available at any given time by the consumer. If they choose to run their air conditioners and not sell energy back into the grid because they need it, it would put the utility in a position where it would have to get the energy from someplace else. This represents a disincentive for the utility to go this route, unless it is able to leverage the demand response programs that are being enabled by AMI (advanced metering infrastructure). The key to distributed generation to include DERs is that the utilities will need the ability to effectively force the consumer to sell them excess energy if it becomes a necessity. The necessity will only come in the event that the utility is required to depend on DERs to maintain generation capacity. Utility companies today rely on the ability to buy and sell energy among themselves through a few thousand generating facilities. Smart distributed generation will require utility companies to buy and sell energy among a disparate consumer base that will include potentially millions of sources from which energy can be purchased.

7.9 CYBER SECURITY AND DISTRIBUTED GENERATION

Historically, generating stations, like the rest of the grid, have been strongly air gapped. In reality, they have no reason to connect to the outside world, as the only thing that truly matters is whether or not they are producing the energy needed to support their customer base. There is, however, some interaction between utilities in regard to how much energy is needed and what is available to sell. This becomes apparent during what are known as load-shedding events. An application within the control center known as automated generation control (AGC) can be used to control generation if it becomes necessary to stabilize the overall system. If a generator fails suddenly, the control center may be forced to shed load or purposely turn off the power to a few consumers in order to ensure that power is sustained to the masses. They may turn up energy from another source using AGC, or they may simply trip circuit breakers to stop the flow of power to certain population bases. Such an event represents the primary concern in relation to cyber security when looking at power generation.

If a cyber security threat were able to suddenly take a large amount of load offline, it would be possible to create a situation where the control center would be forced to shed load from the system. Threats like Stuxnet[29] can even be used to bring down an entire generation facility and stop the output of generation. If the generating facility is large, such as the Three Gorges Dam in China, shedding enough load to compensate for 22,000 megawatts would be extremely troublesome. Keep in mind, however, that the 22,000 megawatts is spread across multiple generating units, as is the setup in most plants. This means that while Stuxnet is a real threat that can cause harm, it may only impact one unit. A plant may have four or five units and one unit going offline may only represent 25 percent of the plant's total generating capacity. This seems to jive with the intent of the NERC CIP (North American Electric Reliability Corporation Critical Infrastructure Protection) standards, to ensure that load-shedding events are limited. To this effect, the NERC CIP standards are intended to prevent a problem with the bulk power system. And by forcing utilities to build to limit load-shedding events to an amount of energy that will shut off power for a few, they can ensure resiliency for the many. This is good risk management practice because it takes into account what the potential effects could be if a threat were able to remove a threat source from service. Assets that can have an impact on the bulk electric system are then required to secure against the possibility that if they were forced to shed their entire load, they could put the grid in a position of cascading failure.

Under the aforementioned circumstance, a cascading failure would likely occur if load is not shed to compensate for the loss of generation, and this would only occur because the control center does not react to the event, meaning that if a limited number of the population is not blacked out because the required amount of energy is not made available, then a cascading failure could ensue. If the supply of energy is not available, then suddenly the demand increases and this strain will eventually lead to a cascading failure.

While Stuxnet could be used to take down a generating facility, it is very unlikely that it could be used to cause an outage to the bulk power system. To destabilize the bulk power system, a threat like Stuxnet would have to be paired with an advanced threat that has compromised systems within the control center and used in conjunction with one another. Because both systems are air gapped, it is unlikely that this would be possible and paired with the timing necessary to pull off such an event. The doomsday scenario would then be where an advanced threat such as an undetected virus or worm gets into a control center system. It then manipulates the data so that the control center is unaware either that a generator is knocked offline or as to the amount of energy that is lost. This would obviously also require that the generator in question actually be kicked offline as well. This level of coordination would likely be necessary because it is very likely that the control center would call the generating facility to determine if the generator failed, validating their need to shed load. If it were reported that a 1,500-megawatt generator went offline and the control center display only showed that 150 megawatts was lost, the impact would likely end up being a cascading failure. This is primarily because no compensation has occurred for the loss of 1,350 megawatts, and that much load would end up being shed simply because the demand could not be met. This might cause confusion in the system among other balancing areas and could result in a serious problem. This potential scenario is meant to illustrate the need for accurate information.

This is really the role that cyber security should play when engaged with the utility community. How do you ensure that the information provided is the correct information? Without accurate information, a scenario such as the one above could occur. Obviously this depends on the defeat of a number of security controls, such as policies preventing users from using thumb drives on the control systems or just good change control. The point is that if it ever did occur, it would not only result in a load-shedding event, but it would also cause system operators to not trust the system. It is this trust that will likely have widespread ripple effects throughout the utility. Operators making decisions using information that cannot be trusted and the fact that those decisions could result in a bulk electric system failure is one of the scenarios that should be studied carefully and guarded against.

As the grid becomes smarter, these decisions are likely going to occur automatically. A system will detect a generator failure and react automatically in order to compensate for the loss. This may mean getting the energy from a demand response solution through interaction with DERs, or it may mean that load will be shed for the few, again to protect the many. But the stakes are even greater in such a scenario as threats no longer need to fool the control center and generation operators; they just need to invalidate the information used by the system to make those automated decisions. Moreover, the fallback measure of simply calling the generation plant to confirm an outage is no longer available as a DER management system run by the utility to balance generation resources may be drawing information from hundreds or perhaps thousands of DERs run by utilities, non-energy businesses, and even residential consumers with a wind turbine in their backyard. This may be done by compromising the DCS system at an individual power plant that is used to feed the control center system used to make these automated decisions, or it may mean a more widespread attack on hundreds of consumer DERs operating across multiple states. If the control center system gets invalid data and makes a load-shedding decision based on that information, the impact could be extreme. This is why there is an emphasis on cyber security when it comes to the modern grid. These types of risks must be analyzed and addressed in order to mitigate the potentially high impact that could ensue.

ENDNOTES

1. For more information, see http://www.aps.com/general_info/about APS_18.html.
2. For more information, see http://www.eia.doe.gov/neic/rankings/plants bycapacity.htm.
3. For more information, see http://www.eia.doe.gov/neic/rankings/plants bycapacity.htm.
4. Matthew Morioka, Alireza Abrishamkar, and Yve Kay, Three Gorges Dam, retrieved March 21, 2011. http://www.eng.hawaii. edu/~panos/444_09_4_9.pdf.
5. http://web.mit.edu/ceepr/www/publications/workingpapers/2009-004. pdf Yangbo Du; John E. Parsons (May 2009) (PDF). *Update on the Cost of Nuclear Power*. Massachusetts Institute of Technology. http:// web.mit.edu/ceepr/www/publications/workingpapers/2009-004.pdf. Retrieved 2009-05-19.
6. For more information, see http://www.nrc.gov/reading-rm/doc-col-lections/fact-sheets/funds-fs.html.
7. Burn plus wet scrubber for exhaust gas cleaning. Crystec Technology Trading GmbH. Retrieved March 22, 2011. http://www.crystec. com/ksiburne.htm.
8. For more information, see http://peswiki.com/index.php/Directory: Cents_Per_Kilowatt-Hour.
9. For a thorough examination of distributed energy resources, see *Federal Energy Management Program Using Distributed Energy Resources, A How-To Guide for Federal Facility Managers,* DOE/ GO-102002-1520, U.S. Department of Energy by the National Renewable Energy Laboratory, May 2002.
10. For a thorough examination of distributed energy resources, see *Federal Energy Management Program Using Distributed Energy Resources, A How-To Guide for Federal Facility Managers,* DOE/ GO-102002-1520, U.S. Department of Energy by the National Renewable Energy Laboratory, May 2002, p. 3.
11. For a thorough examination of distributed energy resources, see *Federal Energy Management Program Using Distributed Energy Resources, A How-To Guide for Federal Facility Managers,* DOE/ GO-102002-1520, U.S. Department of Energy by the National Renewable Energy Laboratory, May 2002.

12. For a thorough examination of distributed energy resources, see *Federal Energy Management Program Using Distributed Energy Resources, A How-To Guide for Federal Facility Managers*, DOE/ GO-102002-1520, U.S. Department of Energy by the National Renewable Energy Laboratory, May 2002.

13. For a thorough examination of distributed energy resources, see *Federal Energy Management Program Using Distributed Energy Resources, A How-To Guide for Federal Facility Managers*, DOE/ GO-102002-1520, U.S. Department of Energy by the National Renewable Energy Laboratory, May 2002.

14. For a thorough examination of distributed energy resources, see *Federal Energy Management Program Using Distributed Energy Resources, A How-To Guide for Federal Facility Managers*, DOE/ GO-102002-1520, U.S. Department of Energy by the National Renewable Energy Laboratory, May 2002.

15. For a thorough examination of distributed energy resources, see *Federal Energy Management Program Using Distributed Energy Resources, A How-To Guide for Federal Facility Managers*, DOE/ GO-102002-1520, U.S. Department of Energy by the National Renewable Energy Laboratory, May 2002, pp. 3–4.

16. See *Federal Energy Management Program Using Distributed Energy Resources, A How-To Guide for Federal Facility Managers*, DOE/ GO-102002-1520, U.S. Department of Energy by the National Renewable Energy Laboratory, May 2002, p. 6.

17. See *Federal Energy Management Program Using Distributed Energy Resources, A How-To Guide for Federal Facility Managers*, DOE/ GO-102002-1520, U.S. Department of Energy by the National Renewable Energy Laboratory, May 2002.

18. See *Federal Energy Management Program Using Distributed Energy Resources, A How-To Guide for Federal Facility Managers*, DOE/ GO-102002-1520, U.S. Department of Energy by the National Renewable Energy Laboratory, May 2002.

19. See *Federal Energy Management Program Using Distributed Energy Resources, A How-To Guide for Federal Facility Managers*, DOE/ GO-102002-1520, U.S. Department of Energy by the National Renewable Energy Laboratory, May 2002, p. 7.

20. See *Federal Energy Management Program Using Distributed Energy Resources, A How-To Guide for Federal Facility Managers*, DOE/ GO-102002-1520, U.S. Department of Energy by the National Renewable Energy Laboratory, May 2002.

21. See *Federal Energy Management Program Using Distributed Energy Resources, A How-To Guide for Federal Facility Managers*, DOE/ GO-102002-1520, U.S. Department of Energy by the National Renewable Energy Laboratory, May 2002.

22. See *Federal Energy Management Program Using Distributed Energy Resources, A How-To Guide for Federal Facility Managers,* DOE/GO-102002-1520, U.S. Department of Energy by the National Renewable Energy Laboratory, May 2002.

23. See *Federal Energy Management Program Using Distributed Energy Resources, A How-To Guide for Federal Facility Managers,* DOE/GO-102002-1520, U.S. Department of Energy by the National Renewable Energy Laboratory, May 2002, p. 6.

24. See *Federal Energy Management Program Using Distributed Energy Resources, A How-To Guide for Federal Facility Managers,* DOE/GO-102002-1520, U.S. Department of Energy by the National Renewable Energy Laboratory, May 2002, p. 7.

25. See *Federal Energy Management Program Using Distributed Energy Resources, A How-To Guide for Federal Facility Managers,* DOE/GO-102002-1520, U.S. Department of Energy by the National Renewable Energy Laboratory, May 2002, p. 3.

26. Chowdhury, S., S.P. Chowdhury, and P. Crossley. *Microgrids and Active Distribution Networks.* The Institution of Engineering and Technology (London), 2009, p. 3.

27. Lyndsey Layton, Reliance on Coal Sullies "Green the Capitol" Effort, *Washington Post,* April 27 (retrieved from http://www.washington-post.com/wp-dyn/content/article/2007/04/20/AR2007042002128. html). The plant uses a mixture of coal, natural gas, and fuel oil to generate energy for the steam and chilled water. Not surprisingly, the local nature of the generation has been a source of controversy both for aesthetic concerns and the pollution that it generates, far above the levels normal for a much larger generation plant.

28. Chowdhury, S., S.P. Chowdhury, and P. Crossley. *Microgrids and Active Distribution Networks.* The Institution of Engineering and Technology (London), 2009, p. 11.

29. Stuxnet was a highly sophisticated attack allegedly targeting an Iranian nuclear plant that was engaged in enriching uranium for possible military purposes. The attack caused centrifuges to spin at frequencies not intended and resulted in the destruction of the centrifuges. Speculation has focused on nation-state actors as the source of this attack. The attack leveraged the use of malicious software that exploited previously unknown Microsoft Windows vulnerabilities. These so-called "zero day" exploits then took advantage of default passwords and known vulnerabilities in a Siemens software package that were used to modify operating instructions for programmable logic controllers and simultaneously make it appear to operators that nothing had changed. Stuxnet is significant not only because of the sophistication of the multi-pronged attack that also likely relied on a significant amount of social engineering, but also

because of the fact that a purely software-based attack was able to cause significant physical damage without the attacker ever directly controlling or even having network connectivity with the targeted asset. More information about Stuxnet can be found at http://www. symantec.com/content/en/us/enterprise/media/security_response/ whitepapers/w32_stuxnet_dossier.pdf.

CHAPTER **8**

Operations and Outsourcing

8.1 INTRODUCTION

Utilities maintain two basic functions: (1) operational management of the grid and (2) support and maintenance of the grid assets. Utilities vary on the organizational structure associated each of these areas, which really depends on the size of the organization. In larger organizations, for example, one will see functions within each of these areas broken down by area of responsibility. A support and maintenance organization within a large utility might have separate line crews to support the varying types of power lines owned, perhaps a line crew to support all the lines that are below 230 kilovolts and a separate group to support power lines larger than 230 kilovolts. Whereas, smaller organizations may do everything in-house, meaning that the line crew may be part of the same group that manages maintenance for transformers. Regardless of how large an organization, a utility will be required to maintain the following support roles:

- Design
- Engineering
- Communications
- Information technology
- Planning
- Grid operations
- Plant operations
- Substation operations
- Accounting
- Marketing
- Substation maintenance
- Generation maintenance
- Construction
- Metering support
- Smart Grid operations

8.2 DESIGN

Utilities must maintain functions that support power system design. These groups serve to design things that need to be built. From grid operations design to building schematics, the design group architects it all. In smaller utilities, this may mean that design is outsourced to a third-party company that specializes in the design and construction of various operational assets. Even in such a case, the utility will need to have someone oversee the design and construction programs. Some examples of what needs to be designed include

- Power plants
- Substations
- Transmission lines
- Electrical circuits
- Engineering

8.3 ENGINEERING

Engineers play an important role in utilities as they are the people who design how things work. There are generally two areas within any given group in a utility organization: engineers and technicians. Engineers design processes and how things will work. For example, if a utility were implementing an IED (intelligent electronic device) into a substation, there would be an engineer within the organization that will design a print to illustrate exactly how and where the IED will be installed. This would include which cables go into which ports, where to plug in the power supply, and other considerations. Technicians are then trained to read engineering specifications and install the devices as instructed by the engineers. There are generally engineers in all parts of a utility organization's operations. They are generally needed to explain how things work and ensure that assets are configured properly.

8.4 COMMUNICATIONS

Utility companies usually maintain a large communications infrastructure that is designed to augment their need to communicate with assets. Remote terminal units (RTUs) require a medium to transmit communications across a wide area. As a result of these remote communications, a group is needed to implement and manage this communications infrastructure. Communications will maintain engineering and technicians, one to design, while the other installs. In some utility organizations, the communications group will also manage the communications devices themselves. Historically, this would be seen as normal. However, these groups do not have the experience with Internet Protocol (IP) communications infrastructures that information technology (IT) groups do. As a result of this experience gap, IT and operations technology are perhaps converging in the area of communications.

8.5 INFORMATION TECHNOLOGY (IT)

IT in utility companies typically supports the enterprise IT infrastructure. These organizations are generally siloed off from the power system organization, which maintains its own separate IT groups. The IT groups within the utility are usually tasked with supporting the control center environment. This is true for the control systems that are usually scattered throughout a utility. At a generation plant, for example, there is a control system that is used to control assets within a power plant. As discussed in previous chapters, the distributed control system (DCS) maintains the ability to control cyber-assets throughout a generation facility. Generation facilities then maintain an IT staff to support the DCS, including the IP communications associated with it. These groups are usually separate from the control center IT environment, which is tasked with supporting the overall transmission and distribution systems' IT support assets. The biggest change for IT that appears to be on the horizon for utility companies is in moving IT into the actual field assets associated with the utility. Field assets refer to substations and other remote locations. It also relates to the concept of substation automation in general. As Smart Grid becomes more of a reality, IP-based communications to substations will increase. While the network communications infrastructure is generally already supported by IT in some fashion, the biggest change will be when the substation infrastructure converts to automated systems such as "Smart RTUs" (RTU = remote terminal unit).

Smart RTUs are basically applications on servers in substations that provide an enhanced RTU capability. What is important to understand with this concept is that IT will be required to support servers in actual substations. These servers will be providing critical field functions in environments for which IT is not traditionally prepared. However, it is widely known that this change is coming, and it may be unlikely that operations will be able to fully support their growing IT needs with existing resources.

8.6 PLANNING

The planning group is responsible for developing plans associated with the future of the power system. For example, planning would determine how much energy the utility company might need in the future and try to figure out how to achieve its energy needs. Planning might decide that it is in the company's best interest to create a new generation source in order to support future demand. This is an important group within the company because their forecasts and plans have a major impact on the future of the organization. If the group does not adequately plan for new generation and the need arises in the future, there might end up being a significant financial impact on the company. This is because the company would have to pay for that new generation from another source. And the cost of future generation might far exceed that of planned future generation needs. The cost of developing new-generation sources now, in order to meet energy needs 5 years into the future, may result in savings and profit to the company. The investment might result in savings because the company may discover that the cost of purchasing energy on the open market in 5 years might be ten times greater than the cost of simply building new generation to begin with. At the same time, the company might discover that they are able to sell that new energy resource at a much higher cost, assuming they find that they didn't need as much energy as anticipated. While these benefits exist, there are also downsides if the planning is off. For example, if the demand for energy decreases because of economic conditions, the company might not have anyone to sell the energy to or may not have a need for it. They might take the opportunity to close older plants that they own in favor of the generation resource.

Renewable energy sources might also play a role in planning, depending on the incentives that the government applies today. If the government provides a cost incentive to build renewable energy resources and to close fossil generation resources, a company might decide to make the investment. In other words, the government can provide enough incentive to make the decision to convert a profitable endeavor where the benefits will be realized in the future. The planning group, however, must determine whether or not the value is there. They might factor in the future perceived political climate and make the change simply because the public wants them to give some incentive. This may allow the company to raise rates in the future under the political cover of the costs associated with the implementation of new renewable energy resources.

The planning group also plans for other new infrastructure that might benefit the company strategically. This may include transmission, distribution, and likely the Smart Grid endeavor. In other words, what is the benefit of implementing grid automation to reduce future costs associated with grid operations? The planning group might take Smart Grid concepts and determine that they can avert major costs in favor of technology solutions. Special protection schemes (SPSs) are just that; they are implemented in order to mitigate the need to undertake a large transmission infrastructure project.[1] The scheme may be necessary to mitigate the risk associated with a secondary path for generation. Whereas a new transmission line might be required to support a secondary path, the line may not be required if an SPS is used instead. This is an example of using a Smart Grid technology to reduce the cost associated with building out additional infrastructure. The planning group will likely make Smart Grid a reality as it is where the real cost-benefit analysis conversation will likely occur within utilities.

8.7 GRID OPERATIONS

Grid operations refers to the monitoring of the power grid. A typical utility with balancing authority responsibilities will establish a control center in order to monitor the stability of their balancing area. The balancing area integrates resource plans ahead of time, maintains load-interchange-generation balance within a balancing area, and supports interconnection frequency in real-time.[2] The responsibility is to basically keep the system in balance within a certain service territory. For example, a balancing authority may maintain a certain area of the system, and those areas are adjacent to other balancing areas. Because these systems are connected to one another, each must maintain balance and coordinate with each other whenever balances are detected. Figure 8.1 illustrates the interconnection between balancing areas, including overlap.

From Figure 8.1, there are three balancing authorities within the West Region. These three balancing areas must interact with one another and coordinate in order to maintain the stability of the west region. West Region balancing area #3, however, is also adjacent to the Central Region balancing area, so they have to maintain coordination among an additional balancing area that the other two do not need to be concerned about. The entire electric system is fashioned in this capacity and it represents the level of coordination necessary to maintain reliability. When

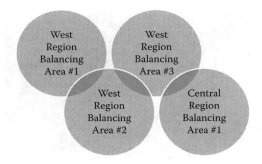

FIGURE 8.1   Balancing authority example.

there are errors in coordination, it can potentially result in grid instability. So when operators in one region or area talk to operators in other regions or areas, they must all be sure of what has, needs to, or will occur.

This area of operations is taken so seriously that NERC requires that their conversations be recorded, per the emergency preparedness and operations (EOP) standards.[3] This means that conversations between operators and technicians are recorded. Operators essentially sit in a command center that somewhat resembles a security operations center (SOC). From this center, the dispatchers can monitor the status of the grid. When they detect line faults, breaker trips, or other anomalies, they can quickly notify the appropriate personnel to go out and correct the problem (See Figure 8.2).

The computer system that supports the dispatch control area is typically managed by a specialized IT group, and commonly referred to as the SCADA (supervisory control and data acquisition) system. These dispatchers are basically the monitoring function for the entire grid. If you take the view that each balancing area maintains a control center and all of the adjacent control area control center dispatchers maintain communications with one another, it becomes easier to understand how the system functions from a holistic perspective. There are a variety of functions that a dispatcher must provide. For example, dispatchers may be required to use a function that sheds load or turns off power to a section of the population. This involves opening a breaker, and they may do this because some monitor or screen has illustrated a disturbance in the system. From a review of the disturbance, the dispatcher follows a procedure to take action that may include notification to the adjacent balancing areas and the use of a decision tree to execute the required checks necessary before taking the action. A smarter grid will help dispatchers, in that the processes that they typically run through will be automated.

Once the processes used by a group are documented and well understood, they can usually be automated. The automation of a process

FIGURE 8.2   Control center example.

represents the highest maturity level that a group can achieve if all the security issues associated with it are properly reconciled. In the dispatcher example above, the dispatcher would first recognize on the map board in the control center that there was a disturbance. As the computer is providing the notification, this process is already automated because it represents the warning system. When the dispatcher receives the warning, he must usually run through a decision tree in order to verify what the warning really is. If this process is mature, then it can be programmed into a system, meaning that the dispatcher will no longer have to run through the decision tree. The system will dictate what the warning means and potentially notify the dispatcher of what he has to do. Or the system may be further enabled to take action, based on the warning and associated definition, to open a breaker somewhere, for example. And it is this fact that makes Smart Grid security relevant.

As has been illustrated in previous chapters, the single biggest achievement in the area of Smart Grid security is in providing the operators the assurance that what the computers are demonstrating to them are, in fact, what they are supposed to be showing. The concern becomes if computers are making all of the decisions, then how would dispatch know that the decisions being made are the correct ones? How would dispatch know that a threat hasn't compromised the system, injected false data, and created a false alarm that was automatically responded to in a manner that is catastrophic? It all then comes back to the integrity issue, and this is where cyber security needs to get involved with dispatch.

Outage detection is another important facet in the management of grid operations. The ability to detect outages is crucial in supporting the reliability of the grid. If an outage is detected, then an operator might be required to take action in order to mitigate the risks of the outage. If a tree falls on a power line, taking it out of service, then the outage detection function might be required to take action in order to limit the outcome of the event. This may mean that they may need to isolate the fault that will occur as a result of the event. This fault isolation may mean that breakers need to be open in order to ensure that the overall system remains stable. If an outage is detected by a system and dispatchers respond to the outage by taking an action to isolate the incident, there will likely be an impact to other parts of the grid. As a result, there must be coordination and collaboration among various departments within the utility, and it is the grid operators who will need to facilitate this coordination.

8.8 PLANT OPERATIONS

Plant operations are similar to grid operations but represent those functions within a distributed process control system. Power plants are really mini balancing areas within themselves, in that they are run by a number of processes. As described in previous chapters, the DCS controls various components within a plant. Without the DCS, there would likely be no remote monitoring or control of the processes that are needed to continue to produce power. Therefore, plant personnel are used to provide the actual monitoring and control of the assets within the plant. The primary function of plant operations is to ensure that the plant remains functional. If the plant ever stops functioning, then the generation resource is unavailable and the company that owns that plant cannot make money or begins to lose it. Therefore, these functions are of great criticality to the operation of the plant. They maintain contact with the control center through various communication media. Control centers even maintain the ability to automatically adjust generation in some circumstances in order to compensate for power loss in other parts of the balancing area. Some power plants are then used to balance the balancing area. As an example, if the control center dispatchers detect the loss of 300 megawatts of load in a given service area, they will likely look to a generation plant in which they have automatic generation control, and increase generation output by 300 megawatts to offset the load loss as previously experienced.

8.9 SUBSTATION OPERATIONS

Substation operations provide the actual field maintenance and communications of the control center to power plants and between balancing areas. Substations typically provide the power systems equivalent to a router in the IT (information technology) world. A router basically routes Internet Protocol (IP) traffic across the network to other routers until the destination is reached. A substation provides a similar function for distribution, transmission, and generation. Power must come out of a plant and into a substation, so that the current can be converted to go across power lines to the next substation. As an example, a 500-kilovolt line may terminate in a substation that splits that amount of power across two 230-kilovolt lines, in order to get 230 kilovolts of power to other substations that may step down voltage and distribute the power to another array of substations.

Depending on the size of the utility's transmission, distribution, and generation infrastructure, a utility can have any number of substations. The Western Area Power Administration's, a Department of Energy Power Marketing Administration organization, Watertown Operations Center owns roughly 9,000 miles of transmission lines that are covered by more than 200 substations.[4] This example is meant to illustrate, by contrast, the number of substations it takes to manage transmission lines. Using the math in the Watertown Operations Center example, there is a substation for every 45 miles of transmission.

There are a number of unmanned substations associated with any utility. And at a ratio of a substation for every 45 miles, it becomes more and more necessary to maintain remote communications between these facilities. In many cases, utilities can run fiber across transmission lines in order to facilitate communications. This communications infrastructure effectively makes utility infrastructures seem like those of an Internet service provider (ISP). In fact, utilities sell fiber access to ISPs such as Verizon and Sprint. These fiber links, however, are maintained by the communications department within substation operations. Some utilities have marketing departments that work to actually sell use of the fiber, but the communications department represents the group that actually manages communications. These groups generally operate at layer one (Physical or Bit Layer) and two (Data Link or Frames Layer) of the Open Systems Interconnection (OSI) model and are generally not engineers with a specialization in Internet Protocol. Information technology usually will augment communications engineering by providing layer three support or Network Layer support. The two groups end up working with additional groups within substation operations in order to provide access to substation assets.

Protection of the power system is also managed out of substation operations. These groups are generally the ones who are tasked with supporting the relays or IEDs that are used to actually trip breakers in the switchyards. These groups are also large in number as they must support devices in each substation that may be more or less than 45 miles from each other. As Smart Grid starts to become a reality, the protection group is starting to implement relays or IEDs that operate via IP. This is increasing the need to involve IT into assets within the field, such as relays. The protection groups really act as the security group for the overall grid, through the assurance that countermeasures will work if any instability is identified. More likely than not, this is the group that should do the most of the collaboration with cyber security teams in order to ensure that their contingency processes are understood so that they can be accounted for during the implementation of cyber security controls. An important thing to understand about the power system is that it is resilient and redundant in many cases. This should tell a cyber security professional that the perceived risk to the infrastructure is lower than it appears. If the grid destabilizes, there are likely countermeasures in place to ensure that it does not fail. Cyber security professionals need to understand what these countermeasures are so that cyber security controls can be designed to augment their the success.

Finally, there is the group within substation operations that monitors and manages RTUs, HMIs, and other devices within the substation.

These groups are typically viewed as the control group, responsible for providing the ability to control IEDs within substations. This group is most closely linked to server and desktop groups within IT organizations. It should be pointed out that the control group is not an IT group and that server and desktop management groups should not be considered groups that can support RTUs and HMIs (human-machine interfaces) within substations. The control group, however, is responsible for establishing RTUs and HMIs that, with the implementation of Smart Grid, are starting to become server-class applications. The control group is responsible for monitoring, measuring, and controlling assets within the substation. Advanced servers can support applications that can process many times more information than the older devices. For example, Smart RTUs may be applications on servers that can provide a number of additional features plus the traditional RTU features to a user. These advanced features may help someone in the control group better support the substation infrastructure.

8.10 ACCOUNTING

The amount of flowing electricity that a utility owns must always be accounted for. The accounting group is basically responsible for ensuring that the amount of energy produced is the amount of energy that may be sold. A generation facility at a utility may, for example, provide 500 megawatts of energy that can be used or sold by the utility. The accounting group understands that 500 megawatts is available when the company may only use 350 megawatts. The result is that 150 megawatts of energy is left over, which represents excess energy. If that same utility ever attempted to use 600 megawatts of the 500 megawatts available, there would be a problem in the amount of energy available. At this point, the utility is forced to purchase 100 megawatts of energy on the open market at an increased price. A utility, however, might also maintain a marketing department to sell excess power.

8.11 MARKETING

Marketing represents the department within operations that markets excess power or access to power. The group monitors what goes on within the power accounting department and determines what they can buy or sell based on what is available. They may choose to purchase energy if accounting shows that enough is not available or they may choose to sell energy if the utility maintains too much energy.

8.12 MAINTENANCE

As stated above, there are various maintenance groups within a utility operations organization. Maintenance usually represents one of the largest budget impacts to the utility because of the many field assets such as lines, transformers, and capacitor banks. Furthermore, these field assets were intended to be in operation over life cycles that exceeded 20 years in many cases. To put it into perspective, if an organization maintains 9,000 miles of transmission lines, what does it take to monitor those lines to ensure that trees don't grow into them and cause disruptions in service, among other things? The answer is through the maintenance and monitoring of the assets themselves. In the transmission line scenario, there are various incidents where trees fall on power lines and cause outages in electricity as a result. Utilities will monitor this by flying over all their transmission lines with a helicopter in order to review the lines for potential vegetation risk. NERC (North American Electric Reliability Corporation) maintains standard FAC 003, "Transmission Vegetation Management Program," which requires utilities with power lines to maintain a program to monitor and maintain power lines.[5] This is an example of the importance of maintenance to the bulk electric system.

One of the benefits of Smart Grid lies in the use of predictive systems analytics that will be used to detect potential faults before they occur. This not only applies to vegetation management, but it also helps solve the problem of lightning strikes and other things that may potentially occur that are prevented by line maintenance. The process will likely work by introducing advanced sensors that can be used to detect the area around line poles. It may also include data from environmental services, including the number of lightning strikes that occur in a given area, as well as how many lightning strikes it takes to cause an outage. If lightning strikes a power line once annually and it takes, let's say, three strikes to take down the line, then the predictive model may set off an alarm in a certain location indicating that the next storm could create a line outage condition. The result could be that preventive maintenance is executed instead of reactive maintenance. These models are far from complete, and there is still ambiguity in how these systems would work, but the benefit is clear as it relates to Smart Grid. Preventive maintenance means higher reliability and less cost incurred by the utility.

8.13 SUBSTATION MAINTENANCE

Transformers and capacitor banks must be maintained for potential issues. An overheating transformer, for instance, can create serious problems in a utility because of the time and cost to replace a bad transformer. The costs associated with transformer failure might include repair or replacement, clean-up of environmental damage, damage to adjacent equipment, lost revenues, and litigation costs.[6] To make matters worse, the aging infrastructure is creating a situation where transformers will likely fail at a rate of 50 percent for those that have been in service for at least 50 years.[7] The detection of such an outage is extremely critical to a utility organization because the cost is significant and the life cycle for a great number of transformers is closely coming to an end. A smarter grid that can help to identify and detect these issues will go a long way in reducing costs to utilities, mostly because detecting a problem with a transformer before it goes bad will alleviate the costs associated with environmental damage, litigation costs, and effects on adjacent equipment. For this reason, maintenance organizations within utilities are very important.

Maintenance to systems within substations such as IEDs is also becoming increasingly important. This is mostly because utilities are accustomed to monitoring and controlling various assets within the switchyard. As the technology is typically older, pressure is increasingly being applied to utilities to replace their older IEDs with IT systems running applications that can simulate what the older technology did and provide better management of this equipment. This is another example of introducing Smart Grid technology to augment maintenance within substations. Information technology follows a replacement life cycle that is different from what is typical of a utility's power assets. One of the issues that utilities will face concerns how to acquire IT professionals on the power systems side in an effort to support these critical field assets.

8.14 GENERATION MAINTENANCE

The maintenance of generating facilities is perhaps the most critical element of a utility. This is because if the generator fails, then the utility is unable to supply power and thus cannot generate revenue. The components of a generation facility are also so large and complex that many times they require vendor support. Maintenance of the IT systems used to augment reliability in a power plant is critical to the support of the plant as well. Many times, DCSs provide monitoring and management capabilities for industrial assets within a plant. If they are not able to provide monitoring of these assets, then failure could be possible without notification. Likewise, if systems become unresponsive from a management perspective, then it becomes possible that action that is supposed to be taken is not taken when it becomes absolutely necessary. The result could be the failure of some assets that can create a cascading maintenance condition.

8.15 CONSTRUCTION

As the energy delivery system expands and Smart Grid technology is introduced, additional construction of new assets and support systems will likely be necessary in order to manage the growth. Going back to the substation control house example, these facilities are many times not equipped to maintain traditional IT systems because of the environmental issues associated with their locations. IT systems require certain thresholds in regard to humidity and temperature to provide assurance that they will not fail. Vendors have been in the process of developing ruggedized IT systems that will survive under these conditions. As substations are in remote locations at times, they are exposed to weather-related conditions. An older substation in the middle of Canada during the winter, for example, may not have the environmental controls to sustain support for a traditional IT system. Construction may then be required to retrofit these facilities so that they can support the IT systems that will improve grid reliability. The construction group within a utility organization is then tasked with establishing structures needed to support assets within a utility company.

8.16 METERING SUPPORT

Meters represent the utility's interface to individual customers. These customers are the revenue base that supports the overall infrastructure. The meters are then the function within the utility organization that is tasked with generating the revenues to pay for everything. Generation exists, for example, so that there is a constant supply of power, which can then be transmitted to a distribution system that will ultimately deliver power to the consumers. The consumers are individual residents and businesses that utilize that power. If the consumer base does not exist, then there is really no reason to generate, transmit, and distribute energy. As a result, meters are the function within the organization that really represents the revenue base for the company.

To utilize energy, one must have a delivery mechanism so that usage can be measured and quantified into terms that can be directly translated to revenue. Power traders, for example, buy and sell energy based on supply and demand. During recessions and bad economic conditions, demand decreases, thereby causing the supply to increase. Whereas, in economic boom times, demand increases, causing the supply to decrease. If consumers are not buying energy or reducing their energy consumption, then less energy is needed to sustain the overall grid. While power is bought and sold at the wholesale level, the ultimate objective is to make it available to a consumer for consumption. The basic premise in the growth of a utility is that the number of people in a service area will increase as the population grows. This population growth translates into new consumers who will require a larger supply of energy. In order to bill consumers for services, meters are needed to measure the amount of usage, which is then quantified into a commodity that is used to determine the price that can be placed on energy consumption.

Traditional meters basically measured energy use and utilities developed mechanisms to acquire this information so that it could be used in billing. With the introduction of smart meters, additional functionality can be added into meters in order to improve the utility's ability to bill and manage power delivery. In earlier chapters, advanced meter reading (AMR) was discussed, representing one of the biggest benefits of smart meters. In traditional meters, utilities were required to maintain a staff of meter readers who were tasked with going out and reading meters so that utilities knew what to bill the consumers. When you start thinking of this in terms of the number of consumers, a large number of meter readers is needed so that the entire service area of a utility can be covered. If there are 3 million consumers, this might translate to 3 million meters. The question then becomes, how many meters can be read by a meter reader in a day, and how many meter reads per day are needed to cover the entire 3 million consumer population in a month? If the answer is that you need 300 readers, reading 3 million meters in a given month at a rate of $20 per hour, then it can be deduced that the financial impact to a utility company in terms of meter readers is $12.5 million a year to support meter readers. AMR allows for these meter reads to occur remotely, which inherently reduces the need to maintain the staff of 300 meter readers. This technology may only require the utility to maintain 30 meter-reading technicians to manage the remote metering reading system. The remote metering system would then save the utility 90 percent of its cost associated with traditional meter reading. The savings realized in such a case represent a cost benefit to the utility because they still get the benefit of the meter reads, but are only required to make an investment in an infrastructure that will yield a large return over the course of time.

There are other benefits to smart metering, including the fact that technicians may no longer be required to drive to a home to disconnect electricity where a consumer has moved from the property. This is because smart meters can now be used to remotely disconnect circuits that supply power to the consumers. There are a number of benefits to smart meters when you begin thinking about the business value associated with remote management of meters. In addition to this, utilities can use the smart meters to modify the behavior of the consumer, thus allowing them to manage power flow through demand response and automated load control.

8.17 SMART GRID OPERATIONS

The implementation of Smart Grids means that the departments that make up operations will have to coordinate in order to achieve success. Utilities that utilize demand response and automated load controlling programs can leverage them to decrease their need to sustain high generation outputs. If the utilities can more accurately distribute and manage the use of power, they may be able to operate their power plants at a lower output level, thereby saving on fuel costs. If the cost of fuel associated with a generation plant that produces 3,000 megawatts of energy is $50 million a year, this represents a direct impact to the utility's cost of providing energy. If the utility could reduce the amount of fuel needed, so that it only needs to produce 2,000 megawatts through demand response and automated load control, then it is possible that it could see the 33 percent savings in its fuel cost and still provide the same level of service to the consumer base. This represents where interaction and the exchange of information between departments within the utility become important. To achieve this level of coordination, both generation and the metering programs are going to have to share information and utilize it in a more coordinated fashion. This may end up being a function that is taken on by the control center, where meter information is made available to generation plants, which represents a reduction in fuel costs.

There are a variety of other opportunities across utility organizations as well that represent the benefits of Smart Grid. It should also be pointed out that there is a significant level of coordination that is already occurring across organizational lines within the utility. The difference is that these processes will increasingly become more and more reliant on information technology for automation. As the processes mature, their automation can be used to achieve a cost benefit to the utility that will likely be directly passed on to the consumer.

The level of automation and coordination represents a unique attack vector for cyber-threats. Smart Grid will require the use of IT to provide the level of automation needed to support the benefits as illustrated above. If the IT systems used to support automation and coordination are not appropriately secured, then it may be possible for cyber-threats to expose weaknesses and cause serious damage to the overall infrastructure. Using the previous example, what happens when generation gets incorrect information from metering and the need for energy is underestimated as a result? The likely result will be an issue in the operation of the grid because incorrect data will be used to make key decisions regarding the appropriate need for power output. In order to mitigate this potential weakness at the strategic layer, utilities will want cyber security professionals monitoring the integration of these systems and implementing controls to reduce the potential risk associated with it.

If there is no malicious code protection or monitoring of malicious code and the system that provides information to the generation plant on how much energy is needed cannot support the current demand, then it is possible that there would be an impact on the metering's Smart Grid programs, such as demand response and automated load control. There might also be an impact on generation, forcing the utility to purchase energy from another entity, which would result in a higher cost and ultimately lost revenue. This is not to say that malicious code protection and monitoring tools will prevent the potential risk, but they do help to provide more assurance that it may not happen. Managing risk at this level will become very important to utilities because it directly supports the mission of the scenario, which is reliability.

8.17.1 Outsourcing

For many years, utilities have relied on the outsourcing of various activities associated with the categories presented above. Outsourcing is extremely important to utilities because they generally either do not maintain enough experience to support the assets in place, or maintain the experience but do not have enough staff to cover the support needs. Outsourcing covers every area of operations within the utility. Design and construction is often outsourced to firms that specialize in the required activities. Vendors are seen as being able to do the job better, quicker, and faster; as a result, they are integral to the utility's operations. The cost of maintaining a full-time staff that can design and construct the asset needs of the organization can be significant. That is not to say that some design and construction does not exist; it simply implies that a hybrid approach exists, depending on what the utility can support.

Outsourcing becomes more critical when it comes to the maintenance and support of the complex systems that are required to operate assets within various areas of the utility. The IT systems used to support the assets are very complicated and require an IT specialization that does not generally exist within corporate IT organizations. The gap in experience is usually attributed to the lack of experience with real-time systems. Real-time systems are generally categorized as high availability systems that must be maintained in real-time. These systems usually cannot be taken offline without incurring an outage to systems operations. An outage means a great deal of coordination and planning, which means that it becomes a process that utilities would like to avoid. In many cases, utilities take scheduled outages to maintain their industrial assets, and it is during these times that outages can be used to upgrade IT systems that support those assets. While the cost of an outage means lost revenue due to repairs and replacements, it is necessary in order to prevent future failures. Unscheduled outages, however, are generally not tolerated. These outage types can also have a significant impact on the bulk electric system because it may require multiple other utilities to react in order to mitigate the risks of the outage. Compliance risks exacerbate this in that they are regulated by NERC and FERC many times. If the energy management system (EMS) fails due to the implementation of a patch, the risks mentioned above can be quickly realized.

Vendors have taken action in recent years to manage this risk by requiring utilities to gain approval from the vendors before patches and other IT maintenance issues are performed. The vendor many times will manage the risk of the implementation of patches and other IT maintenance needs by executing testing on its own end. Some vendors are even providing their own applications that have direct feeds to the vendors that update the utilities in real-time on what they can and cannot do from an IT maintenance perspective. For example, if a patch is released for the Microsoft Windows platform that is used to support the real-time system, the vendor may test and implement the patch on its systems in order to determine the scope of its impact. If the patch does not create a problem for the vendor, then it becomes unlikely that it will be a problem for the utility. This is basically a risk management practice designed to reduce the probability that a utility will have to take an unscheduled outage. And this is the primary source of concern as associated with the gap in experience related to IT and operations technology (OT).

High availability means something in an IT organization, but real-time means something slightly different to a utility. The complexity of these underlying support systems also plays a factor in this, which is why outsourcing of support is often required. Cyber security represents an additional risk to the utility because of the actions and functions required to mitigate cyber-risk. This creates concern in the industry because cyber security is not well understood. Furthermore, there have not been any reported significant cyber-incidents caused by a malicious source, which makes utilities view the implementation of cyber security controls as a bigger risk than what the threats might indicate. From a cyber security professional's perspective, the threats are going to increase over time, and they see a very vulnerable system. The cyber-risk to the systems from a cyber security perspective is that a lack of security controls would likely result in the utility's systems being compromised or taken offline easily. And from the utility's perspective, these systems are tightly air gapped, which represents a significant reduction in risk. However, with the use of outsourcing, there are various access vectors that could be taken advantage of by threats being the vendors themselves. The questions then become: how secure are the vendors, and does the risk increase as a result of outsourcing?

8.17.2 Cyber Security Incident Response and Outsourcing

The current view of risk to a utility's power system is seen from a siloed approach. The risk of a compromise appears to be low because systems are air gapped from the corporate network, which maintains the external connections to the Internet. When you look at the same systems from an enterprise perspective, you quickly see that there are lots of attack vectors through direct connections to various vendors and service providers. This means that the opportunity to inflict damage exists apart from the traditional connections to the Internet. Control systems do not directly communicate with the Internet in most cases, but vendors do maintain direct connections to control systems for maintenance purposes. Utilities many times do not understand the cyber security posture of their vendors and service providers, and therefore maintain a trust relationship. But have these vendors been evaluated for cyber security? And if something were to happen on the vendor side, would they make that information available to the utility? It is possible that vendors have a sterling cyber security posture and are all working to protect their clients. However, it is also possible that they have been or will be compromised and are subjecting their clients to cyber-risk. The point is that if you do not evaluate the security posture of the vendor, then it is possible that the risk is unknown and therefore goes unmanaged.

Monitoring for cyber-threats through an incident identification and response strategy should then extend beyond the traditional boundaries of the utility itself. Agreements must be in place for notification if something were to occur. While coordination and collaboration for maintenance purposes is a good thing, it is hard to know what the security posture of all the maintenance and service vendors is. Vendors are typically connected to multiple utilities that are connected to multiple vendors, which means that the number of attack vectors into a utility could potentially be limitless. And the question becomes: if vendor A is compromised, how many utilities does it affect? And how would those utilities know if they were affected or not?

To mitigate this risk, utilities and vendors must begin to insert cyber security into their maintenance and support contracts; this is known as vendor management. If a vendor loses information deemed to be private, then they are generally required to report the fact that their was a breach and that this type of information was lost or stolen. However there appears to be no such legal requirements for a vendor who is compromised and who has direct access to a utilities control system. This is important because a compromised vendor could result in a compromised utility and disclosure of the attack to the utility will allow them to potentially develop remediation plans to mitigate the risk of compromise at the utility as a result of what happened with the vendor.This is as opposed to many privacy rules that require credit card companies and others to notify an organization if they have had a breach. This is a result of privacy information being treated as being more critical than the control systems themselves. As a part of a good incident response security posture, this sort of collaboration may be necessary in the highly interconnected organizations that support the bulk electric system, including utilities, vendors, and service providers. Furthermore, rules and regulations that require utilities, vendors, and service providers to notify each other and the regulatory authority in the case of a cyber security incident might represent a mechanism that is used to enhance incident response capabilities for the power industry.

NERC maintains a cyber security threat and vulnerability program that serves to analyze vulnerabilities associated with industrial control systems. Utilities are required to report incidents that occur to Sector Information Sharing and Analysis Centers (ES-ISAC) NERC ES-ISAC program within NERC; however, reporting of an incident could result in sanctions or even an investigation of the utility that made the report. This naturally causes some resistance to reporting because of the perceived relationship. And this is a large issue within the industry that various groups and organizations are trying to solve. The creation of an organization with no perceived regulatory oversight would likely garner more reporting from utilities. More reporting from utilities would likely result in better threat intelligence that would help in identifying solutions to mitigate the risks identified. The key problem today is, given the complexity of the infrastructure, all the interconnections between organizations, and a lack of history regarding cyber-threats to the power system, that the risk is neither well defined or nor well understood. There are effectively no data that one can point to showing that a cyber security threat has exploited a utility control system.

Stuxnet and Night Dragon are examples of cyber-threats that were used to exploit control systems. However, these threats were very targeted and specific to the systems that they compromised. In the case of Stuxnet, the malware was essentially brought in on a thumb drive with the clear intention of bringing down an Iranian nuclear power plant. It is difficult to visualize such a threat to utility companies in the United States, given all of the compensating controls surrounding those infrastructures: homeland security, police departments, background checks, cyber security monitoring tools, and other resources. As a result of all these controls, the threat is really limited to targeted threats. Targeted threats are those individuals, groups, or even nation-states that wish to cause specific harm to a specific utility or service area.

If a targeted threat wanted to compromise a utility company or the power system assets themselves, then they could simply engage an internal employee, pay them, and gain the desired access. This is usually the way it works in the intelligence community, so why would it be any different in a utility? The question then becomes: is it worth the effort to gain such access? What is the cost-benefit of hijacking critical infrastructure and causing power disruptions? And then the reality of the situation is that there is no money to be made by a traditional threat in doing this. Compromising a control system at a utility is then an act of war. And this is a war that utilities are not prepared to fight, again because of the cost-benefit. Attempting to track down an advanced targeted threat on a network using traditional security controls is almost impossible. A good hacker will be able to get into the system and will probably never be detected. A utility might waste millions of dollars looking for the advanced threat and never find it. The control system might be shut down and the affected utility will likely undergo a series of regulatory investigations so that the root cause can be established.

The liability of a security compromise on a utility control system can only be managed. Implementing cyber security controls to prevent common threats that illustrate that a utility organization was not actually negligent during an investigation would likely mitigate a lot of risk. As a result, a utility might implement a standard set of access controls, identification and authentication, system integrity controls, system communications controls, and audit logging, and reporting tools to identify and respond to incidents that occur. The accepted model of protection would then be a defense-in-depth infrastructure that prevents and detects malicious activity enough to demonstrate that a utility is not negligent in its cyber security protection activities. This means that the air gapped nature of the systems is not enough to illustrate that the utility was not negligent. The utility needs to demonstrate that it understands the different attack vectors (access points) into its control system and that it has placed layered defenses around those access points in order to mitigate risk. As part of the layered approach, the utility also needs to implement controls to prevent and detect events that would occur if its access points were ever breached. Essentially, security to prevent intrusions and security to detect intrusions if their prevention mechanisms work are required. NERC CIP provides a good framework for getting this type of infrastructure in place for the utility; the problem is that it only addresses part of the problem.

NERC CIP attempts to prioritize risk through the development of the risk-based assessment methodology that requires utilities to identify critical assets and associated critical cyber-assets. This, however, only accounts for a percentage of the utility infrastructure and basically does not do anything to require cyber security on the remainder of the infrastructure. Furthermore, the NERC CIP framework only addresses the bulk electric system as being the "thing" that could be impacted. There is no methodology for identifying critical assets within the service territories. Hospitals and military bases, for example, are not evaluated and therefore cyber security controls are not required to protect those assets. In any case, it would be up to the utility to implement basic layered controls in order to reduce their overall liabilities in the event that something does occur.

8.17.3 Cyber Security Controls

Access controls obviously need to be applied at the network perimeter for utility control systems. Access from corporate environments into operations environments should be brokered by a series of access points where specific access is defined. This basically ensures that only ports and services required for the operation of the control system asset are able to access the control system asset. As an example, a user who must access an IED within a substation should be made to go through a specific system type (perhaps a terminal interface) in order to access the IED. The access point should only allow the terminal interface access to the IED, and the user must pass through this control point in order to gain access.

The terminal interface also provides the ability to record who the users are and some assurance that they are who they say they are through the required username and password. Layering this identification and authentication security on top of the access point security provides two layers of defense. If a threat attempted to access the access point bypassing the terminal interface, then access would be denied because the access point only allows for terminal interface access. The threat would then have to attempt to compromise the terminal interface itself and would be required to identify and authenticate itself in order to access the system. If the threat does not have a username and password, then it would likely be denied access. The effectiveness of both controls may lead a threat to attempt unconventional attack methods, and this is where the threat begins to get sophisticated. For example, a threat may attempt to spoof the IP address of the terminal interface knowing that it maintains access through the access point. In doing so, a control must be in place to detect spoofing; usually a firewall used as an access point can serve this purpose.

If the spoofing attack fails, the threat will likely result to using attack tools and techniques to gain unauthorized access to the terminal interface. This means that they will attempt to start taking advantage of the flaws in the operating system and applications, which represent the terminal interface. If the operating system that supports the terminal interface is not patched, for example, a threat might be able to use the weakness to gain what is called "backdoor access" to the terminal interface. Backdoor access is where the flaw in the system that the patch is meant to correct can be used by a threat to access it by bypassing the identification and authentication module. There are tools widely available on the Internet that provide this capability. The primary mechanism to correct problems like this is to maintain good patch management processes to ensure that systems are updated in a timely manner, effectively reducing the amount of time a threat has to execute such an attack. The longer the systems go unpatched, the more time a threat has to take advantage of a weakness. There are a number of controls available to prevent access. The utility companies need to simply figure out which combination offers them the best balance with regard to management of risk.

Utilities are increasingly seeing the need to maintain cyber security professionals within their organizations. The establishment of a team to manage cyber-risk to critical infrastructure is then becoming an important element in the protection scheme for the power industry. Maintaining a qualified cyber security professional reduces the risks associated with cyber-threats because a utility will at least be paying attention to them. The point is that the organization needs to have someone monitoring the potential for a cyber security incident, identifying risks in processes and procedures used to manage the system, and developing policies to reduce risk that may exist.

ENDNOTES

1. Smart RAS (Remedial Action Scheme). Shimo Wang, Senior Member, IEEE George Rodriguez, Member, IEEE http://asset. sce.com/Documents/Environment%20-%20Smart%20Grid/ SmartRemedialActionScheme.pdf.
2. For more information on NERC, see NERC Glossary at http://www. nerc.com/files/Glossary_12Feb08.pdf.
3. For more information on NERC EOP Standards, see "NERC EOP" at http://www.nerc.com/files/EOP-002-3.pdf.
4. For more information on Western Area Power Administration Watertown Operations, see http://www.wapa.gov/newsroom/ WTbrochure.htm.
5. For more information on NERC FAC Standards, see "NERC FAC" at http://www.nerc.com/files/FAC-003-1.pdf.
6. http://www.bplglobal.net/eng/knowledge-center/download. aspx?id=196, page 12.
7. http://www.bplglobal.net/eng/knowledge-center/download. aspx?id=196, page 6.

CHAPTER 9

Plug-In Electric Vehicles and Energy Storage
Now the Fun Really Begins

9.1 INTRODUCTION

As noted in Chapter 7, the success of unreliable distributed energy resources such as solar and wind will heavily depend on the ability to store the energy for times when the wind is not blowing or the sun is not shining. For example, according to a paper to be released by the National Solar Energy Center in Israel, it concluded that "[s]olar plants, without storage, could only generate around 3 percent to 4 percent of the country's power without being forced to dump large amounts of solar power."[1] That figure goes up substantially if the solar plants are paired with energy storage. The quest for reliable and cost-effective energy storage lies front and center in our effort to both make effective use of renewable energy and satisfy the needs of our increasingly mobile society. It is also central to the successful deployment of electric vehicles where consumers, accustomed to driving 300–400 miles on a single gasoline fill-up, are frequently balking at the suggestions of a 5- to 8-hour charging cycle every 50 miles or so. And beyond consumer adoption is the logistical challenge of deploying vehicle charging stations in a manner where driving all-electric vehicles could be practical.

More importantly, for the Smart Grid, plug-in electric vehicles represent a broader technology known as energy storage. This comprises traditional battery technologies found in gas-powered vehicles, golf carts, and various indoor-use vehicles, toys, and other devices that require a small amount of energy for a short duration. When pressed for higher demand uses, battery technology has proven to be a disappointment, with so-called "power users" resorting to gas-powered motors or electrical outlets. It seems that as the number of mobile devices proliferates, the more we learn to appreciate the need for reliable delivery of electric power as laptop and mobile phone users scurry about in airports and conference rooms looking for the nearest available outlet. As network communication becomes nearly ubiquitous, our biggest obstacle to mobility appears to be power. Interestingly, the ability to provide effective, efficient, and affordable energy storage is also one of the key enablers of the Smart Grid, particularly distributed generation and renewable resources.

9.2 STORAGE TECHNOLOGIES

As just noted, energy storage is not new. It exists in nearly every electrical device we use, from a backup function for temporary power outages to those designed for mobility where frequent re-charges are necessary. Even the traditional electrical grid makes heavy use of energy through capacitor banks usually located at substations that tend to be designed more to balance the flow of electricity rather than to provide an alternate power source. Additionally, energy can be stored in more natural ways. For example, pumped storage is a relatively old technology that takes advantage of gravity and electricity demand variances to cost effectively provide power. Under most scenarios, pumped storage, such as the system depicted in Figure 9.1, stores water in a large reservoir in an elevated location.

When electricity is needed, water is released and flows downhill through hydroelectric turbines much like a hydroelectric dam. However,

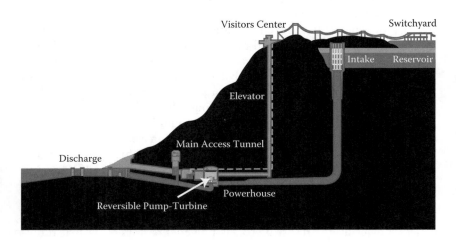

FIGURE 9.1   Taum Sauk pumped storage reservoir. (From U.S. Geological Survey, USGS/Rolla, Missouri.)

unlike a river that naturally replenishes the reservoir, the water that is released is collected in a second reservoir at a lower elevation and pumped back up during times of low energy demand. In theory, the same principle could be applied to renewable energy, where solar and wind power could be used to do the pumping. Unfortunately, pumped storage has not proven very popular as it requires a rather large footprint, is costly to construct if both reservoirs do not already occur naturally, and requires a fair amount of maintenance.[2]

Other "natural" technologies face similar drawbacks. For example, utilities have looked into channeling energy into compressed air either in conjunction with wind turbines or as a completely independent function. The compressed air could be stored in underground caverns and converted back to energy when needed. Similarly, on a smaller scale, building operators are experimenting with technology that generates ice during periods of low energy demand and then uses it at peak times to provide cooling.[3] Nearly every physical and chemical process includes the notion of potential energy, and so the possibilities are endless. However, the practical realities dictating cost effectiveness and low environmental impact tend to significantly limit options. Moreover, the need for flexibility and standardization pretty much calls for chemical-based solutions typically found in battery technology. For years, scientists have been researching different chemical properties to determine which ones are most effective at storing and releasing energy over extended periods of charge and discharge. So far, the results have been somewhat unimpressive, with plug-in vehicles being one of the prime examples. Additionally, the higher cost of the vehicle usually more than cancels out the lower energy costs related to charging. However, significant investment is targeting storage technology to hopefully make dramatic improvements.[4]

One example of energy storage for larger-scale use is the semi-trailer sized batteries that are being deployed in pilot programs. One such product is the TransFlow 2000 sold by Premium Power. "It is a fully integrated system that comprises energy storage, power conditioning, system control and thermal management subsystems packaged into a portable, turn-key, building block to be placed wherever it is needed for immediately dispatchable on-line energy storage. Each TransFlow 2000 provides up to 500kW of power and 2.8MWh of energy storage capacity in a single enclosure that fits onto a 53' trailer for mobility."[5] These storage units can be transported around, depending on the need, or could be dedicated to a particular geography. Some have suggested the notion of neighborhood-based storage that could blunt the effects of a storm or give utilities more granular options when generation capacity is pressed to its limits. However, these storage technologies are not cheap and are relatively unproven in production environments for extended periods of time.

9.3 MEASUREMENT AND COORDINATION

Additionally, people who have seen their laptop's battery life unpredict-ably shorten after repeated use understand the importance of effective instrumentation to anticipate the amount of power remaining when driv-ing an electric-only vehicle or drawing from battery storage to power a critical appliance in the home. In the past, most homes and businesses received their power from a single power plant that was typically dic-tated by geography and did not change over time. With more transmis-sion lines and greater diversity in generation resources, the power grid offered greater resiliency and redundancy. However, that required more coordination by regional entities such as regional transmission opera-tors (RTOs) and independent system operators (ISOs), as described in previous chapters. However, unlike these regional entities that typically coordinate dozens or perhaps hundreds of generation assets, a wide deployment of localized energy storage resources at the distribution level requires an extensive amount of coordination that must be done by an automated system. Telephone communication just won't work. This is even truer for plug-in electric vehicles that could potentially be lever-aged by the power grid as a storage resource. Instrumentation built into storage resources must accurately provide data on the amount of energy capacity remaining as well as predict the amount of energy required by local demand. For example, someone may choose to sell some of his car's stored energy during peak times of the work day as long as there is enough remaining to get the owner home in the evening. Additionally, each storage resource must be interoperable with the grid's communica-tion network and effectively authenticate itself, so systems attempting to balance energy understand the unit's underlying capabilities.

Within this context, cyber security is critically important. As has often been repeated, a breach of integrity is often the equivalent of directly controlling the device as bad information leads to poor deci-sions and actions that may be unwise. For example, large batteries are typically filled with toxic and caustic chemicals that are harmful to the environment and people if they overheat and catch on fire.[6] Proper maintenance and measurement of the electric charging and discharging process is crucial. Because battery storage is likely to be widely distrib-uted, a communications infrastructure is needed to both ensure that the devices are properly maintained and also to instantly be aware of the amount of charge available to address demand spikes. Incorrect readings could lead to power outages when loads are shifted to battery power that is not there. Moreover, the sophisticated calculations must be made that determine how much load the batteries can handle and for how long. Just like distributed generation, energy storage requires a very precise balance. At the moment, most battery-based storage is not centrally dispatchable. While it may be paired locally with solar and wind resources, the ability to be truly useful for the larger grid will require connection to a communications infrastructure that would allow a utility to automate decisions about whether the storage device would be used. This is inherently where the "rubber meets the road" in a Smart Grid context. Utilities simply cannot count on small-scale energy storage and distributed generation that is not reliably dispatchable. It may make for good public relations that people can sell back power to electric companies, but in the aggregate, it does little good for the local utility if it excessively relies on that energy resource to be there for the next 100-degree day only to find out that all those net energy producers out there went back to being net energy consumers. It is only through accurate instrumentation of these power and storage resources and a robust communications infrastructure that utilities will be able to forgo the building of new generation plants or the use of more costly "peaker" generation plants.

9.4 WHAT MAKES PLUG-IN ELECTRIC VEHICLES UNIQUE?

At first blush, it would appear that plug-in electric vehicles (PEVs) are no different than any other kind of electric storage. After all, in many cases, the same battery technologies are used. However, calling PEVs just a kind of energy storage is characterizing a smart phone as no different than a land-line phone. PEVs have many more features and are likely to be adopted to a much greater extent than typical energy storage. While typical battery storage may be associated with a distributed generation source, it is just as likely to receive all its charging from the main electrical grid. With hybrid plug-in electric vehicles (HPEVs), there is the option of charging the battery through the use of an internal combustion engine normally powered by gasoline. It remains to be seen whether consumers will choose all-electric models or hybrid models, or both. Nonetheless, it is likely that hybrid models will be in use in the near term. Next, an electric vehicle is a mobile energy resource. A utility cannot count on the storage resource to be in a predictable location or even available at all. Moreover, "[e]lectric or plug-in hybrid vehicles represent a substantial electric load in comparison to standard household appliances. If PHEVs penetrate the market in volumes necessary to reap the projected benefits, they will have to be considered in the load forecasting and distribution system considerations of utilities."[7] Additionally, if utilities are going to go to the effort to incorporate these vehicles onto the grid, it would also make sense that efforts are made to incentivize their use to limit charging during high demand periods and to leverage their energy storage capability during those same windows. However, doing so is extraordinarily complicated for a device that is still in its infancy both technologically and with respect to consumer adoption. Nonetheless, it is imperative that the consequences of PEVs as a load and as a source of stored energy for the grid both be considered together. Failing to do so could either lead to utilities building too many new or too few generation resources to deal with the additional demand. For example, it is highly likely that some percentage of PEVs will require charging during peak demand periods. If that demand is not offset by the stored energy from the vast majority of PEVs that do their charging during off-peak hours, more generation will likely be needed during those peak times.

9.5 PLUG-IN VEHICLE TO GRID LOGISTICS

Now that we've established the relative uniqueness of electric vehicles as an energy storage resource and future significant load, it is worth examining exactly what utilities and consumers need to do to bring these devices onto the grid. To begin with, manufacturers need to make these vehicles grid-aware and be able to engage in two-way communication. Moreover, if an electric vehicle is ever going to be a storage source for the grid, utilities must have the ability to collect and analyze sophisticated data measurements provided through the electric vehicle supply equipment (EVSE), discussed in Chapter 4, working in conjunction with the battery management system that may be integrated with the vehicle and manage the battery pack that powers the vehicle. "The battery pack management techniques are general and suitable for any energy storage-based system, such as electric and hybrid vehicles, distributed power generation units, and portable consumer electronics. Of all the applications, the most rigorous usage of energy storage systems is in hybrid vehicles where it goes through pulsed charge/discharge cycles. Hence, battery pack management is required to be of the most advanced type."[8] Such a management system "is designed to have all or some of the following features:

- State-of-charge (SoC) estimation
- State-of-health (SoH) monitoring for cell and pack protection
- Temperature control
- Charge/discharge power control
- Cell equalization
- Data logging"[9]

This data should be accessed by the utility and analyzed across an entire fleet of electric vehicles that are available to provide power to the grid. Moreover, one could also assume that the charging process would also leverage this data in combination with data on time-of-use pricing and any demand response events that may have occurred. Finally, the vehicle's owner will need to input his or her driving plans, so that the vehicle can determine the amount of energy needed for the vehicle and when. For example, if a mother knows she needs to take the kids to soccer practice, pick up a gallon of milk and some produce at the grocery store, and pick up the dry cleaning, an on-board global positioning system (GPS) could be used to select those locations and thereby calculate the distance that will be traveled, the average speed, and other considerations in determining the power needed. Of course, that same mother may get a last-minute request to swing by the mall after soccer practice to check out the latest toys. To be successful, the algorithms should take all of that into accounts and perhaps engage in a little behavioral profiling. Consequently, there will undoubtedly be some significant privacy concerns that will need to be addressed. Additionally, utilities will need to find ways to incentivize consumers to make their stored energy available, and manufacturers will need to supply vehicles with the ability to communicate battery status information and consumer preferences to the EVSE and then over the HAN (home area network) or some other network. For example, one article estimates that a PHEV owner could earn from $184 to $3,285 per year selling back excess energy to the grid.[10] At the upper end, that could potentially pay for the energy used for charging the car in the first place. However, it is likely that the amount would be at the lower end, thus making it difficult to persuade consumers to go through the trouble. However, the good news with PHEVs is that if the vehicle owner's plans change, the gasoline engine can be used to make up for any lack of battery power. Moreover, if utilities want to emulate the automatic generation control (AGC) functionality that is available from traditional generation sources that provide spinning reserve, then utilities will need to gather information in real-time. Typically, AGC signals are sent every few seconds with the reserves having to be available within 10 minutes of a contingency arising.[11] In some cases, utilities may have more advance notice of a shortage in generation capacity, but it would seem that battery power would be ideally suited for rapid response. The real question will be: is it worth the challenges in dealing with such a highly decentralized storage resource that may be connected to the grid one moment and disconnected the next with little or no notice?

It should come as little surprise then that cyber security will likely be important. Like distributed generation, vehicle owners are getting paid for the energy they provide to the grid, making it critical that the energy source is properly identified so that the proper account can be credited. Additionally, utilities will generally only need this energy during peak periods and will need to incentivize the price appropriately. The system will need to accurately record the time of day that the energy is supplied. Consequently, cyber security controls will need to ensure that hackers cannot send false identifiers or manipulate the time or date. There may also be a competition for who is able to supply the utility with energy during peak times when the price is highest. To ensure fairness, cyber security controls will need to prevent manipulation of the selection system used by the utility. Similarly, utilities will want to operate some sort of mini-balancing authority for electric vehicles to ensure that energy is received on the grid in phase and at the right frequency. In addition to an attacker intentionally trying to cause harm to the grid, utilities will need to guard against malfunctioning vehicles. In most cases, the amount of disruption from a single vehicle is unlikely to cause any harm. However, utilities will want to watch closely the introduction of vehicles onto the grid, as unanticipated conditions always seem to pop up at the worst possible times. As noted previously, strict privacy controls must also be implemented that define who can access data on a customer's travel plans as well as his or her usage history. Finally, the vehicle and the EVSE components that are communicating with the utility must be protected from other vehicle owners through HAN security controls (discussed in Chapter 4).

9.6 GRID TO PLUG-IN VEHICLE LOGISTICS

Regardless of whether utilities will be able to figure out a plan to leverage plug-in electric vehicles to inject power into the grid, they will have to deal with electric vehicles demanding power from the grid. Introducing this new load on a widespread basis has profound implications for today's electrical grid. For example, a 50-percent market penetration of PHEVs in the Northeast corresponds to about a 10-percent increase in overall electricity demand.[12] Assuming that the vast majority of charging is done during non-peak hours—usually late evening or early morning hours—the grid should have the capacity to handle that. "This type of new load represents an opportunity for the electric utility industry to expand sales without contributing to system peak."[13] While the generation capacity may be capable of handling the additional load, the infrastructure in between the generation and the vehicle may be less able to adapt. Aging transformers that are now able to "rest" during the night could potentially be stressed 24 hours a day and possibly lead to more failures. Additionally, people's homes may need retrofitting to support faster charging to both limit the amount of time that the vehicle's owner is unable to use the vehicle and to ensure that the entire charging can occur outside the peak usage windows.[14] Unfortunately, utilities will have little control over when these vehicles are introduced to the grid. Plug-in vehicles exist today that can be connected to a standard 120-volt outlet and with the help of a local electrician, a 240-volt outlet is easy to install without any involvement of the utility, assuming that the home has sufficient capacity to support it, a likely scenario if other major appliances are not being used at the same time. However, government regulations may dictate that the vehicle be metered separately to both give the utility the flexibility to develop programs specifically for electric vehicles and for taxing authorities to collect needed revenue for roads and related expenses. The latter issue is starting to get attention as federal, state, and local jurisdictions realize that without action on their part, the tax revenue collected through a gasoline tax will disappear if electric vehicle owners are getting their energy from the electric company rather than the gas pumps. Because electric vehicle owners have the same impact on roads as those with gasoline-only engines, it would make sense that taxes should be incurred equally on both groups. Those with their own solar or wind power may be able to escape the tax altogether, but that group is likely to be small for the foreseeable future.

That raises the issue of payment. With the exception of taxes, it would seem reasonable just to consider an electric vehicle as just another appliance with no need to segregate its charging cost. However, the vehicle is the only major appliance that moves, and it will likely need to be charged when away from home. Some have proposed a billing system that would treat cars like cell phones and allow an owner to charge his vehicle wherever he chooses and be billed by his electric company based on a unique identifier for the vehicle. However, that scenario is not likely to be realistic. Instead, the owners of the charging stations, whether it be a private residence or a commercial charging facility, will likely be billed for the charge and will need to implement their own method of recovering the costs of the charge independent of the utility, but would obtain the current rates from the utility. A private residence would likely just estimate the cost while a commercial charging station would likely use up-to-the-minute pricing information from the utility and operate just like a gas station.[15]

While, in theory, the risks to the grid of vehicle charging are not much different from those for electricity supplied to any battery-based appliance (such as a laptop computer), the potential energy demand is on a much higher scale. Consequently, fraud and potential disruptions to grid operations are much more likely. The simplest fraud would be to make an electric vehicle appear to be something else. At a minimum, that method could potentially allow someone to escape tax liability; and based upon the fact that without the vehicle-to-grid communication, a utility would find it difficult to know for sure that a vehicle was being charged. As Figure 9.2 demonstrates, there may be ways that slight

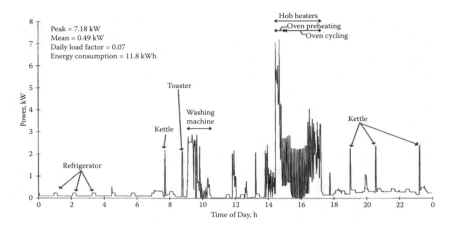

FIGURE 9.2   Energy usage by appliance.[16]

differences in how appliances use electricity can be leveraged to identify a vehicle, the newness of the technology would seem to imply that this will be challenging.

Moreover, the fact that such data are available may give people pause not only from a privacy perspective, but also from possible manipulation by parties seeking to promote their own agendas. In essence, we could be fighting the same battle being fought over net neutrality as electric companies may increasingly be able to selectively deliver power to certain appliances in the same way that power is often restored first to emergency services after an outage. Using load control technology, utilities could decide which appliances are essential and which are not during a time of peak usage. People could find their televisions working but their iPods unable to charge. Such a possibility could generate an underground market of hacking tools and impersonation software designed to make a particular device appear to be one deemed essential. The fact that electric vehicles will likely create an increased demand that utilities may need to respond to with device-based pricing programs and preferential treatment of certain devices during partial outages certainly provides motivation for manipulation.

9.7 ENERGY STORAGE AND CYBER SECURITY

As we've seen, cyber security is an important ingredient in any energy storage technology. When providing stored energy to the grid, a utility must not only be able to process data and queries that it expects, but it must also be able to handle the unexpected. With such a massive increase in storage, the traditional means of relying on a human being to perform a sanity check just won't work. The systems must be designed to reject data outside a permissible range and possibly bar that device from participating in any storage-to-grid program until the right ranges are provided. This helps to ensure that malicious or inadvertent bad data do not cause any cascading effects. For example, if an electric vehicle reports that it has ten times more energy available than an electric vehicle is capable of storing, then the utility should preclude that device from providing power to the grid until that anomaly is corrected. Because the communications data are so inextricably linked to the electric power, anomalies on one should raise suspicion for the other. Moreover, with the potential for so many energy storage resources being available, utilities can afford to be picky. With respect to grid to storage, the security challenges are less complex, but it is nonetheless important that any communication received from the storage device or similar unit be carefully scrutinized. We are now adding a new class of devices with significantly greater energy demand, and individuals will seek to get as much power as quickly as possible in order to shorten the charging window, particularly for electric vehicles. Despite the fact that the average vehicle spends 90 percent of its time parked, drivers are often insistent that their cars be ready at a moment's notice. That may give some the incentive to accelerate the charging process by manipulating the amount of power coming into their homes by either sending the utility false information or other actions that could tax the power generation capability and potentially cause harm to people and property. Taking measurements at various points in the grid and reconciling those measurements is one way to remain vigilant to possible abuses.

Additionally, utilities should consider some sort of certification program that is tied to the storage device's cryptographic key and an embedded identifier before the utility is willing to receive electricity from the device. For electric vehicles, the EVSE is what communicates with the utility. However, because electric vehicles may use multiple charging units, it may still make sense for the vehicle to have its own identifier and cryptographic key, so that the utility can certify the vehicle itself and block vehicles deemed harmful. This may also help in giving vehicle owners credit for the energy provided to the grid if that business model is chosen. Otherwise, a public charging station may need some way to credit the vehicle owner, such as through reduced or free parking or to an account already established. Clearly, the industry has a long way to go before there is seamless interoperability across charging stations that will allow vehicle-to-grid solutions to be truly viable and dispatchable. Successfully implementing such a system may be almost as complex as deploying the original electric grid. In some sense, it may be more complex given the two-way nature of the communications and the tremendous amount of coordination that needs to occur. The security and interoperability challenges alone are likely to delay adoption for a number of years.

9.8 THE FUTURE OF ENERGY STORAGE

Energy storage is likely to be a key ingredient in future Smart Grid deployments. Given the amount of wind and solar being deployed, it makes sense that storage be used to even out the volatility of those technologies. More importantly, electric vehicles present an option for energy storage that is too great to pass up. Because its role as an energy storage resource is secondary to its primary purpose of human mobility, its capital cost to the utility is virtually nothing and much less than a dedicated energy storage resource. The challenge will be in organizing all those decentralized resources into something equating to a power pool. In order to do so, cyber security will be key to any solution.

ENDNOTES

1. Michael Kanellos, Can a country get 90 percent of its power from renewables? *Wired*, April 3, 2011, retrieved from http://www.wired.com/epicenter/2011/04/country-90pct-renewables/.
2. This was proven very true after the Taum Sauk pumped storage reservoir shown in Figure 9.1 experienced a breach of one of its reservoir walls due to lax maintenance and inaccurate water-level measurements coming from sensors that had dislodged from their moorings. This caused millions of dollars in damage as billions of gallons of water gushed onto nearby land, nearly causing the deaths of several individuals. It took approximately 5 years for the facility to return to operation. For more information on the breach and the ensuing investigation, see http://www.ferc.gov/industries/hydropower/safety/projects/taum-sauk.asp.
3. One example of this is the Ice Bear Energy Storage System from Ice Energy. "Daytime energy demand from air conditioning—typically 40–50% of a building's electricity use during peak daytime hours—can be reduced significantly. In kilowatts, each Ice Bear delivers an average reduction of 7.2 kW of source equivalent peak demand for a minimum of 6 hours daily, shifting 32 kW-hours of on-peak energy to off-peak hours." http://www.ice-energy.com/ice-bear-energy-storage-system.

4. See Katie Fehrenbacher, 5 Energy storage players that won Smart Grid stimulus funds. *Gigom,* November 24, 2009, retrieved from http://gigaom.com/cleantech/5-energy-storage-players-that-won-smart-grid-stimulus-funds/ (noting "that energy storage technology received 16 grants for a total of $185 million from the U.S. federal government").

5. Retrieved from http://www.premiumpower.com/product/transflow2000.php.

6. As example of how problematic energy storage can become, a community in the Florida Keys known as No Name Key that is completely disconnected from the grid and rely on solar power, diesel generators, and battery storage, is opting to connect to the grid, with some arguing that a grid-powered option is greener than the fumes generated by diesel generators and the dangers posed by acid batteries. See "No Name Key—In the News," retrieved from http://www.nonamekey.org/nnknews.html.

7. U.S. Dept. of Energy, Plug-in Hybrid Electric Vehicle R&D Plan, http://www1.eere.energy.gov/vehiclesandfules/pdfs/program/phev_rd_plan_02-28-07.pdf, p. 32.

8. Husain, Iqbal. *Electric and Hybrid Vehicles: Design Fundamentals,* 2nd ed., CRC Press, Boca Raton, FL, 2011, p. 159.

9. U.S. Dept. of Energy, Plug-in Hybrid Electric Vehicle R&D Plan, http://www1.eere.energy.gov/vehiclesandfules/pdfs/program/phev_rd_plan_02-28-07.pdf, p. 32.

10. Husain, Iqbal. *Electric and Hybrid Vehicles: Design Fundamentals,* 2nd ed., CRC Press, Boca Raton, FL, 2011, p. 160.

11. See Steven Letendre, Paul Denholm, Peter Lilienthal, Electric and hybrid cars, new load or new resource?, *Public Utilities Fortnightly,* December 2006, p. 28.

12. See Steven Letendre, Paul Denholm, Peter Lilienthal, Electric and hybrid cars, new load or new resource?, *Public Utilities Fortnightly,* December 2006, p. 31 (noting that "[t]hese reserves are used only when a scheduled generator trips offline or a transmission or distribution facility fails, and must be up to full power within 10 minutes.").

13. See Steven Letendre, Paul Denholm, Peter Lilienthal, Electric and hybrid cars, new load or new resource?, *Public Utilities Fortnightly,* December 2006, p. 31.

14. See Steven Letendre, Paul Denholm, Peter Lilienthal, Electric and hybrid cars, new load or new resource?, *Public Utilities Fortnightly,* December 2006, p. 30.

15. "For residential or private and most public charging locations, there are two power levels: Level I and Level II. Level I or convenience charging, allows for charging the traction battery pack

while the vehicle is connected to a 120 V, 15 A branch circuit. A complete charging cycle takes anywhere from 10 to 15 hours to be completed. This type of charging system uses the common ground electrical outlets and is used when Level II charging is unavailable. Level II charging takes place while the vehicle is connected to a 240 V, 40 A circuit, dedicated solely for EV traction battery charging purposes only. At the Level II voltage and current levels, a full charge takes from 3 to 8 hours, depending on battery type. In order to sustain the Level II power requirements, EVSE must be hardwired to the premises wiring." Dhameja, Sandeep. *Electric Vehicle Battery Systems*. Newnes, Boston, MA, 2002, p. 88.

16. In fact, commercial charging stations may offer what are known as Level III charging systems. "Level III charging is defined as the EV equivalent of a commercial gasoline service station. In this case, a Level III charging station can successfully charge an EV in a matter of minutes. To accomplish Level III charging, the equipment must be rated at power levels from 75 to 150 kW. The Level III requires supply circuit to the equipment be rated at 480 V, 3Φ and between 90 to 250 A. However, the supply circuit for the Level III charge may be even larger in capacity. The equipment is to be handled by specially trained personnel." Dhameja, Sandeep. *Electric Vehicle Battery Systems*. Newnes, Boston, MA, 2002, p. 88.

CHAPTER 10

What about the Consumer?
Securing Relationships between the Utilities and Their Customers

10.1 INTRODUCTION

Consumers are the most important aspect of any utility company, as they are the primary source of revenue. Consumers exist at multiple levels within a utility, as there are residential, business, and other utilities that all represent the consumer base. Residential consumers are probably the most visible consumer, as they are the direct recipients of power from the distribution system at the smart meter level. An example of a residential consumer is a family of four living in a home using energy in support of their lifestyle. Business customers are similar to residential customers, but likely use more power because of their size. An example of a business customer would be a shopping mall or office building that utilizes energy in order to sustain a business. Finally, energy is bought and sold in the transmission and generation systems between utilities. Utilities may pay to utilize another utility's transmission line in order to clear a path so that the former can provide energy to a certain area. And energy is bought and sold between utilities at the generation level. These models represent the traditional view of the consumer to a utility; however, that view is being expanded because of Smart Grid.

Smart Grid is introducing new methods of energy delivery to consumers. In the distribution and smart meter space, for example, electric charging stations, home area networks, and distributed generation, are being introduced as methods of acquiring and controlling energy use in a way that the general public may not have believed was possible. The introduction of these forms of energy delivery provides an opportunity for consumers to become more involved in the management of energy.

10.2 ELECTRIC CHARGING STATIONS

Electric charging stations will allow the utility to forge a relationship with consumers that they normally would not have access to. Charging stations will likely become the gas stations of the future, in that they will be accessible by anyone. Charging stations will not only be used for the purchase of energy for electric vehicles, but might also be used to charge anything that requires energy. The point is that consumers who are not current utility customers will end up purchasing energy from the utility. This will effectively expand the potential consumer base for utility companies. For this reason, the utilities will have additional incentive to ensure that charging stations remain reliable. As consumers will likely utilize charging stations with their credit cards and other forms of electronic payments, consumer trust will become increasingly important.

The use of payment centers for electric charging stations will also add complexity to the utilities, in that they may find themselves bound to Payment Card Industry cyber security requirements, among other standards. Furthermore, these stations will be unmanned and with virtually no physical security protection. This physical exposure makes them targets for thieves and other threats. Physical exposure is a risk because cyber-threats will likely utilize the exposure to attempt to interface with them electronically. The implementation of such stations will then have to encompass electronic protection that mitigates the risk that cyber-threats will be able to interface with them and steal customer information, including credit cards. Without a cyber security approach for this, electric charging stations will be at significant risk. Moreover, considering the number of electric charging stations that will be deployed and the extensive communications infrastructure necessary to support them, the risk becomes extremely heightened without a cyber security strategy. To take advantage of the benefits that these stations will provide, the utilities will have to provide assurance to the consumer that they are safe to use.

10.3 HOME AREA NETWORKS

The home area network (HAN) will likely become the primary method that the consumer utilizes to communicate with the utility. This may include an interface that not only allows them to monitor their energy usage in real-time remotely, but to also interact with the utility. As a result, the HAN offers the utility a unique ability to tie the utility communications network to the user's home network. To improve and sustain good customer relations, the HAN interface in the home might be used to communicate important messages to consumers. For example, instead of sending them an electric bill or providing a mechanism to review their bill on the utility's website, there may simply be a meter on the HAN interface that demonstrates this information in real-time. The utility might use the HAN interface to communicate messages to consumers about how they can improve their energy usage. When there are cyber security issues with the utility, messages regarding it, including precautions, would likely be presented on the HAN interface. Or the utility might use the HAN interface to broadcast cyber security awareness information, as the issue will become more and more important to the consumer as Smart Grid ties into their life directly with the use of energy. Another use may be in emergency communications, where the consumer receives emergency messages issued by the government. The opportunity to build trust and confidence with the consumer is probably going to be seen as one of the most effective ways to augment customer service.

10.4 DISTRIBUTED GENERATION

An increasing number of consumers are already utilizing generation in their own homes and selling it back to the grid. This process engages the consumer on a level that starts to really illustrate the benefits of a smarter grid. The utilities might allow the consumer to participate in the marketing process of buying and selling energy at the distribution level. This might be represented by a consumer looking at an interface that tells them the current need and responding by selling their energy output directly to a buyer, with the utility facilitating the transaction. As a result, the benefits of maintaining independent generation at the consumer level become increasingly beneficial. The utilities that need it will potentially be able to purchase energy at a lower cost, as the consumers would simply be selling excess energy that they do not need. This will allow the distribution systems to operate more efficiently and not require the purchase of wholesale energy. This will ultimately reduce reliance on the bulk electric system, effectively balancing out the risks between both distribution and generation.

10.5 DEMAND RESPONSE AND THE CONSUMER

Demand response is sometimes seen as a threat by consumers as it is one of the smart metering programs that will allow the utility to come into the consumer's home and turn off components in order to stabilize the overall system. The consumer will likely have the ability to opt in and opt out of demand response programs, unless there is a state of emergency. This will require the utility to carefully craft a trust relationship so that the consumers will buy into a utility's need to potentially disrupt service in order to correct a problem that it might have been perceived to have caused. Again, trust and confidence will need to be instilled in the consumer in order to make the programs work as intended. There are also cyber security implications with such a program if the utility network is compromised by a threat and the threat exercises the demand response program in an unauthorized manner. Utilities must be prepared to address issues such as this.

10.6 CONSUMER HEALTH RISKS OF SMART GRID

There is a perceived risk to human health as a result of the radio communications signals sent and received by smart meters. This fear is associated with the belief that RF (radio frequency) electromagnetic fields can cause health problems, including cancer. The Federal Communications Commission is required by federal law to study the effects of RF electromagnetic fields on humans. As result, various studies have been conducted over the years that have been used to determine the frequency ranges that would be acceptable to human health. The American National Standards Institute (ANSI), the Institute of Electrical and Electronics Engineers, Inc. (IEEE), and the National Council on Radiation Protection and Measurements (NCRP) have all issued recommendations for human exposure to RF electromagnetic fields.[1] As a result of these recommendations and studies performed by the FCC (Federal Communications Commission), it has been found that the risk to human health depends on the frequency associated with the RF signal.

The NCRP recommended that the Maximum Permissible Exposure limits for human health be in the 300-kilohertz to 100-gigahertz frequency range. Anything above this frequency could then be considered a potential risk, but the study shows that anything within this range is not a risk to human health. Furthermore, because smart meters operate within these ranges, they are not considered a threat to human health. Utility companies will have to utilize facts and figures such as these in order to ensure that the consumer is aware that they are operating within safe boundaries with regard to RF. The key is in the socialization to the consumer that smart meters are safe.

10.7 CONSUMER PROTECTIONS

Consumer protections are an important aspect of Smart Grid, because they are in place to help the consumer. As such, there are concerns on the part of consumers in relation to the implementation of Smart Grid capabilities. Perhaps the greatest concern is in the accuracy of billing associated with smart meters. Traditional meter reading was originally based on physical reads of electromechanical meters. As population bases grew and the numbers of consumers that had to be managed by utilities increased, meter reading became based on periodic reads of meters and estimates in order to identify the correct usage. Smart meters provide accurate reads of meters on demand—and up to the minute in some cases. This causes consumers to maintain higher electric bills, citing that the accuracy of the smart meters is causing higher bills for consumers. Because utility bills might be higher, consumers may choose to seek protection from what might be perceived as unfair billing practices. Recently, controversy erupted over the accuracy of new smart meters in California[2] and Texas.[3] Ultimately, only a very small percentage of the meters were found to be inaccurate, with the remainder of the complaints being attributed to the weather and more accurate measurement of energy usage. Nonetheless, it demonstrated why utilities need to work harder to engage their customers and dispel any false rumors or other misplaced concerns.

As a result of these issues, consumer advocacy groups are lobbying to ensure that certain issues are addressed as Smart Grid becomes a reality.[4] Some of these proposals include

- Requiring that utilities share the risks associated with the deployment of smart meters, not allowing them to pass the potential losses on to the consumer through rate increases.
- Ensuring that dynamic pricing cannot be made mandatory, in order to prevent utilities from forcing consumers into price plans that might end up costing them more money.
- Asking regulators to analyze alternatives to smart meters to determine if there is an alternative to load management, so that utilities will be able to provide a choice to the consumer with regard to balancing the system.
- Requiring utilities that utilize smart meter programs such as remote disconnect, to ensure that they utilize it in a manner that allows the consumer to be treated fairly. Consumers should continue having the right to contest bills and disputer problems and issues before they are disconnected.
- Requiring that utilities address privacy and cyber security concerns before smart meters are rolled out.
- Requiring utilities to engage in consumer education to ensure that consumers maintain the appropriate level of understanding so that they can make an educated risk-based decision.
- Ensuring that Smart Grid investments are verified through transparency to ensure that utilities can be held accountable for the costs incurred and the benefits that are supposed to be obtained as a result of Smart Grid.

10.8 UTILITY PROTECTION FROM THE CONSUMER

As we saw in Chapter 4 in our discussion of HANs, consumers are in control. Utilities need to assume that data coming from HAN devices could have been modified or injected with malicious content. The information supplied should be viewed suspiciously and not relied upon without some corroboration. However, the bigger danger for utilities is the interface they provide for direct interaction with the consumer. For decades, utilities' only interaction with the consumer was the bill that was sent in the mail. Occasionally, which usually means during outages, consumers would also interact with customer service personnel to obtain status information or to make another inquiry about their service. However, these calls were usually not positive experiences. The call was motivated by a problem and because thousands of others were usually having the same problem and therefore had to wait for someone to assist them, the interaction was hardly a positive one. Through the use of technology, however, utilities are trying to change that relationship. Internet-based portals have been deployed to first support online bill payment and are now rolling out services related to the Smart Grid components that we've been discussing. This includes a portal for consumers to track their energy usage that is typically linked with a customer information system that may receive direct feeds from the meter data management system discussed in Chapter 3. The capabilities of these portals may vary, depending on the progress of a utility's Smart Grid program. Typically, the portal provides usage data in the form of graphs, bar charts, and other representations to help the customer understand the reason(s) for the size of his or her bill. The danger for the utility is that the portal can give customers and those pretending to be customers an avenue of attack that could possibly lead to a compromise of more critical systems because of the relationship between the customer information system and others within a utility. For example, the diagram depicted in Figure 10.1 shows a simplified version of the how the consumer accesses the utility's portal and how data flows back and forth between the portal and other utility systems.

Typically, the customer is given access to the customer portal, which queries the customer information system for energy usage and billing data. An application in between may massage the data and deliver it to the portal for the customer to see. In a legacy portal prior to AMI (advanced metering infrastructure), this is effectively where the automated data flow would end. In the past, data collected from meters in the field would have to be manually loaded onto the customer information system. However, under AMI, the process is automated. Either the customer information system pulls or the meter data management system (MDMS) pushes meter data for the customer information system

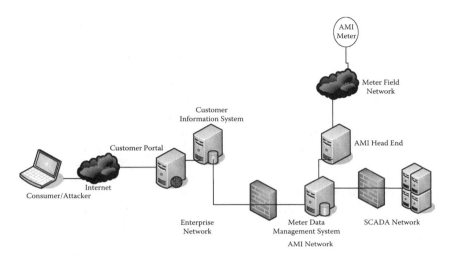

FIGURE 10.1 Customer's Internet interface to utility.

to consume. Similarly, the MDMS receives meter data from the AMI head-end system, which obtains data from the meters, often with the assistance of a collector as described in Chapter 3. Additionally, the distribution SCADA (supervisory control and data acquisition) system discussed in Chapter 5 may also want access to the meter data stored in the MDMS. All this would be relatively low risk if not for the fact that data and commands can also flow in the other direction. For example, the operator of the customer information system can initiate a request to issue a disconnect request to a meter. The request would go to the MDMS, which would instruct the AMI head-end to issue the instruction to the meter. Consequently, despite the very clear separation between the AMI and enterprise networks, data and commands are flowing between and introducing risks that the data and associated commands will be unauthorized. In the immediate example, an attacker posing as a customer could attempt to elevate his access rights in order to initiate a request to a group of meters to disconnect electricity. In many cases, an injection attack on top of a legitimate data query can be used to run an unauthorized program at privilege level of the application handling the query. "For example, a SQL Injection query will be executed with the privileges of the application querying the database. Because many applications are configured with administrator level access to one or more databases, the SQL Injection query will be associated with administrator level access...."[5]

The likelihood that exploiting such a vulnerability in the database would lead to a disconnect order being carried out will depend on what other defenses are implemented. For example, the customer portal may be located on a system that is isolated from the AMI functions and simply has read-only access to the customer database. In that case, the biggest danger would be for an attacker to elevate his permissions in order to view other customers' records. While that is a significant privacy concern, it is certainly less severe than being able to shut off electricity to thousands of homes. In general, it is good practice to examine the potential impacts if an attacker were to gain administrative access. If those impacts are severe, then it is wise to consider segregating systems. However, under those circumstances, attackers can often gain access to one system and leverage that access to another. Consequently, firewalls, intrusion detection systems, and other protections should be considered. Additionally, business rules can be hard wired to limit the impact of an attack. For example, if a utility does not desire to issue disconnect requests to more than a handful of meters during a particular time period, then the application should not present that as an option and systems downstream, such as the head-end system, could also implement restrictions on the number of meters they can issue disconnect commands to in a given time period.

Beyond the customer portal, a utility's other major concern is what an individual consumer may do to the meter and other devices in the HAN. In Chapters 3 and 4, we covered some of the AMI and HAN threats, particularly those coming from a dedicated attacker. However, there are also concerns about what the average customer can do with little or no effort. At the top of the list of concerns is energy theft. Right now the challenge is manageable. With older mechanical meters, customers could rig the meter's dial to move more slowly. But that problem is manageable because most customers would not go through that effort. However, with networked meters, the danger is that an exploit that manipulates meters to report less than the actual energy usage could spread quickly and be used by thousands of customers. In some countries, this has already happened with some assistance from corrupt utility employees.

10.9 THIRD-PARTY SERVICE PROVIDERS

As previously discussed, the Smart Grid is not just the province of utilities and their customers. Increasingly, others are joining the effort to capitalize on this grid evolution. Among them are third-party service providers that are supporting both utilities and consumers directly. This includes analytics services such as Google's PowerMeter, which is "a free energy monitoring tool that helps you save energy and money. Using energy information provided by utility smart meters and energy monitoring devices, Google PowerMeter enables you to view your home's energy consumption from anywhere online."[6] The online tool requires either coordination with a participating utility or a compatible device like the TED 5000. "The Energy Detective (TED) [from Energy, Inc.] is an accurate home electricity monitor that provides real-time feedback on electricity usage."[7] Typically, devices such as the TED 5000 collect power readings from key junction points in the home and then transmit them to a local display and/or to an application such as Google PowerMeter to be analyzed and incorporated into the information displayed by PowerMeter. Similarly, GridGlo, a data analytics software developer and service provider, is offering utilities tools and services to analyze meter data in combination with publicly available information such as financial records and satellite imagery. The objective is to better forecast demand for power. "GridGlo's plan is to sell applications to utilities or access to its data. Over time, it expects to make that data available to third-party companies to create custom applications. For example, a developer could write an application that links credit card reward programs to a utility energy-saving program...."[8] These types of opportunities are likely to make privacy advocates a little queasy. However, the promise of Smart Grid can only be realized if data collected can be analyzed and leveraged appropriately. Historically, utilities have not been particularly skilled at analyzing large chunks of data. The resources available have often been cumbersome and for a limited purpose, such as providing aggregate totals of energy consumed for a particular region. Like it or not, marketing will be key to consumer adoption of Smart Grid programs, and data analytics is key to successful marketing. Moreover, utilities have a valuable asset in the usage data they collect. While privacy concerns must be addressed, anonymized versions of usage data can be very valuable. For example, the fact that a utility could determine what appliance is being used when and for how long, as depicted Figure 9.2, would be incredibly valuable for appliance manufacturers. Going one step further, if consumers agreed, appliance manufacturers could offer a service to owners of their products that would alert consumers when their appliances were not being used in an energy-efficient manner and possibly recommend when a replacement is needed. That may seem creepy to some, but for some segments of the population, it could be a valuable money-saving resource. Plus, manufacturers could afford to give these customers fairly significant discounts in exchange for the data the customers are consenting to provide.

Additionally, many utilities will likely need to outsource much of the operational analytics, such as demand response programs that require massive amounts of data to be analyzed in real-time to determine the right combination of generation increases and load shedding in order to maintain balance. Companies such as SAIC (Science Applications International Corporation) and BPL Global (Better Power Lines Global) are now offering these services to utilities with meters and other data flowing up to the service provider for analysis and then the recommended actions being returned to the utility for execution. Some are labeling this concept "Smart Grid as a Service," which is reminiscent of the cloud computing monikers of software-as-a-service, platform-as-a-service, and infrastructure-as-a-service. Moreover, this is just the tip of the iceberg. Utilities are struggling to deal with aging workforce issues and often find it difficult for existing staff accustomed to a more mechanical operation to support these new technologies. It is likely that Smart Grid will be used as a test case to determine what other services to outsource. Cyber security is one possibility, through the managed security services that are now offered. The exact model for how this would be implemented for Smart Grid is still being ironed out, but the business proposition is compelling. One small- to medium-sized investor-owned utility is a perfect example of this need. The utility has about 150 employees located in a single location and has about 100,000 customers. For the individual overseeing the security of 150 desktops and a smattering of servers in the enterprise, the job is fairly manageable. However, dealing with alerts, even false ones, from 100,000 meters will likely be overwhelming. That is where a managed service that leverages economies of scale over dozens or even hundreds of customers could be valuable.

Of course, cyber security poses some additional challenges when new players are introduced into the equation. All the marketing data flowing out of the utility must be sanitized of any identifier that could be traced back to a particular customer. Moreover, even sanitized data can often present clues as to the identity of customers. For example, power usage correlated with a pool pump could be associated with a particular customer if that residence is the only one with a pool in a particular geographic area. Of greater consequence is the number of third parties that will need access to parts of the utility's infrastructure. In some cases, one-way data feeds may be sufficient. Those flows can be made reasonably secure, particularly where the data has been anonymized. However, many of these functions are inherently two-way. A service provider supporting the utility's demand response operations will need detailed usage data, and will need to send commands and possibly directly control the control system and metering equipment. In these situations, utilities are often under strict guidance. At a minimum, utilities may need to implement two-factor authentication, closely monitor the traffic for any anomalies, require the service provider's personnel to undergo background investigations and security training, and meet the other requirements of the utility's security policy, which may include compliance with NERC CIP (North American Electric Reliability Corporation Critical Infrastructure Protection). Additionally, that usually means that the service provider will need to provide documented evidence of all compliance for the utility to use in support of regulatory audits. For this reason, it is important that service provider activities are sufficiently segregated to only the systems, data, and networks that they require access to. By drawing the lines correctly, the utility may be able to escape audits for the service provider functions. Nonetheless, it is still important from a security risk perspective that appropriate controls are in place, as potential impacts can be large if an attacker were able to send out incorrect demand response signals to meters or generation.

In addition to the standard security controls used by utilities, most will need to implement an expanded vendor management program that takes a more holistic view of security, tracks background checks and training, and conducts regular audits. For service provider functions covered by NERC CIP, this will be critical as it has slowly found its way into other industries that regulate information, such as health care through the Health Insurance Portability and Accountability Act (HIPAA), financial services through the Gramm-Leach-Bliley Act (GLBA), and government through the Federal Information Security Management Act (FISMA). Each of these laws and the supporting regulations has led to the creation of vendor management programs to ensure that vendors are meeting the relevant regulations. In the case of HIPAA, enhancements to the legislation included in the Health Information Technology for Economic and Clinical Health (HITECH) Act directly impose the information security obligation on service providers. In Canada, the Personal Information Protection and Electronic Documents Act (PIPEDA) imposes an obligation on all organizations to protect personal information, and the European Data Protection Directive and supporting laws and regulations of member states impose similar obligations.[9] At a minimum, utilities should develop questionnaires to be answered by all service providers that will be handling customer information, directly or indirectly operate control systems owned by the utility, or provide guidance to how such systems should be operated. Such items would include

- Are background checks done for all employees accessing utility data?
- Is information security training provided annually?
- Is there a written information security policy that identifies all roles and responsibilities related to security for your organization?
- Are data on laptops and mobile media encrypted when not in use?
- Are access authorizations documented and access rights implemented only after written authorization?
- Have software developers been training on secure software development practices?

- Are disaster recovery and business continuity procedures documented and tested on at least an annual basis?
- Is two-factor authentication required for all remote access (e.g., over the Internet)?
- Please identify network security protection deployed (e.g., firewalls, intrusion detection systems, host intrusion prevention systems, Web proxies, virtual private networks, spam filters, data leak prevention) and how they are deployed.
- Describe how you protect against malicious software (e.g., antivirus software, application whitelisting software, restrictions on administrative rights).
- Will services performed for the utility be done in a secured facility on networks that are segregated from the rest of the organization's network infrastructure?

10.10 PROTECTING CONSUMERS FROM THEMSELVES

One of the biggest challenges that utilities face when dealing with consumers and the Smart Grid is ensuring that consumers do not incur self-inflicted wounds. Without appropriate guidance, consumers will buy home automation devices that can be easily hacked, or worse. For example, power measurement tools such as the TED 5000 must be connected to the home junction box where the power enters the house from the distribution feeder line. With the current averaging about 200 amperes, doing a poor wiring job could lead to a fire or electrocution. Similarly, electric vehicles may draw more power than any other appliance in the home and can pose similar dangers if not treated with care. In many cases, safe operation depends on having accurate information about the state of a device. An electronic display that shows the electricity turned off can be deadly if that information is wrong. While less harmful but more frequent are the situations where consumers are tricked into downloading malicious software that disables HAN devices or simply compromises their home computer with promises like "save 50% on your energy by installing this software." It is important that utilities provide guidance and, where possible, offer packaged solutions that are vetted by the utility for interoperability and security. Ideally, the utility should also take responsibility for maintaining these devices for a monthly fee. However, many consumers will want to be independent or don't have enough trust in the utility. One of the biggest challenges utilities face with Smart Grid rollouts is building that trust with consumers. For some utilities, the wrath incurred as a result of unexplained outages, long restoration times, indecipherable bills, rude customer service staff, and unpredictable field support is not easily overcome. However, there are many utilities around the world that have provided world-class service and are rewarded with customers who are quicker to follow security instructions, participate in energy efficiency programs, and pay their bills. The Smart Grid is not a "do-over," as customers will still remember how they were treated in the past, but it can be an opportunity to reengage and, over time, restore some of that trust. For security to be implemented correctly in all parts of the Smart Grid, this trust will certainly be needed.

ENDNOTES

1. See Klaus Bender, No Health Threat from Smart Meters, retrieved from http://www.smartgridnews.com/artman/uploads/1/NoSmartMeterHealthThreat_1_.pdf.
2. See Katie Fehrenbacher, Lesson Learned from the PG&E Smart Meter Suit: It's a Communication Problem, November 19, 2009, retrieved from http://gigaom.com/cleantech/lesson-learned-from-the-pge-smart-meter-suit-its-a-communication-problem/.
3. Katie Fehrenbacher, Smart Meter Backlash, Again: This Time in Texas, March 10, 2010, retrieved from http://gigaom.com/cleantech/smart-meter-backlash-again-this-time-in-texas/.
4. See The Need for Essential Consumer Protections: Smart Metering Proposals and the Move to Time-Based Pricing, AARP, National Consumer Law Center, National Association of State UtilityConsumer Advocates, Consumers Union, and Public Citizen, August 2010. http://www.nclc.org/images/pdf/energy_utility_telecom/additional_resources/adv_meter_protection_report.pdf.
5. Tony Flick and Justin Morehouse, *Securing the Smart Grid: Next Generation Power Grid Security*, Syngress, Burlington, MA, 2011, p. 128.
6. Retrieved from http://www.google.com/powermeter/about/about.html.
7. Retrieved from http://www.theenergydetective.com/about-ted.
8. Martin LaMonica, Start-up GridGlo Taps Smart-Meter Data Deluge, CNet News, May 11, 2011, retrieved from http://news.cnet.com/8301-11128_3-20061749-54.html?part=rss&subj=news&tag=2547-1_3-0-20.

CHAPTER 11

Identifying and Recovering the Grid from a Cyber-Disaster

11.1 INTRODUCTION

Perhaps the biggest concern with regard to Smart Grid security lies in how incidents will be identified and, more importantly, recovered from after their occurrence. This is because cyber security incidents will have a direct impact on the reliability of Smart Grid. Moreover, as time progresses and the Smart Grid starts to become a reality, we will likely bear witness to cyber security incidents that will impact the Smart Grid. This is primarily due to the increasing number of interconnections and the use of information technology (IT) assets in this space. As the grid transitions from the use of traditional communications infrastructures to IP communications infrastructures, the risk of an incident will increase. Without appropriate security controls, these IP infrastructures will be exposed to cyber security threats that have historically been viewed as issues that have had no relevance in this space. An increase in IP (Internet Protocol)-based infrastructures will likely result in an increase in IT involvement in the automation of processes designed to augment the Smart Grid.

As discussed in previous chapters, more connections to vendors and other utilities increase the risk exposure because utilities will rely on these entities for cyber security risk mitigation. Furthermore, we will likely see an increase in the reduction of the "air gap" that exists between operations and corporate data networks because of the need to use the IT infrastructure to support the OT (operations technology) infrastructure that will become increasingly reliant on IP-based communications. To mitigate these risks, utilities will see an increased need to develop security solutions and protocols to protect their IP-based operations networks. As operations is in the business of operating grid assets, it are not in the business of supporting large IT infrastructures that will be needed to operate those assets. The question then becomes: how will incidents be identified? How will operations know that threats exist on their IP networks and on their IT assets?

It is clear that the threat that wishes to exploit a weakness in an operations network is likely to cause instability issues. This is because the mission of the operations network is to support reliability, and the value to a threat would be to cause instability. If a threat exists on an operations network, it will need to be identified, contained, and eradicated in short order. There are two basic threat types in such an environment:

1. Malicious threats
2. Nonmalicious threats

Malicious threats are those that wish to intentionally cause a problem, while nonmalicious threats are those that are unintentional.

11.2 MALICIOUS THREATS

Malicious threats are those that represent the intent to do harm. These threats fall within multiple categories but can be summarized as those that are seeking to do harm. They are probably the most dangerous threat types because they are there to intentionally create an undesired impact. A malicious threat might want to manipulate information within an operations environment so that operators will make an unintentional mistake. As illustrated in previous examples, threats that change information so that it misrepresents what an operator expects are likely scenarios that will have a significant impact. For example, if a threat manipulates an IT system that reports the status of a line to illustrate that there is a fault when there is no fault, it may cause an operator to open a breaker when it is not necessary. It may even cause an automated system to automatically open a breaker when it is not needed. As most operators will tell you, this is a problem because opening a breaker will close a circuit, and cause them to do additional things to compensate for the outage. In extreme cases, if someone is working on the line that is supposed to be de-energized when in fact it is energized, someone will likely be killed.

Malicious threats take various forms, but the probability of impact is seen as low. This is because of the air gapped nature of utility control systems. Utilities are careful to isolate their systems from corporate data networks and from the Internet. Upon review, one will find that there is usually a firewall or some restrictive access point that only allows authorized ports and services through to an operational network. The operational network is where substation operations are usually managed. These are usually systems that are required to communicate with field devices for maintenance purposes. And finally there are the control systems networks such as SCADA (supervisory control and data acquisition) that are isolated from the operational network. The isolation is done by access points, usually firewalls, in order to set a demarcation point as to where the network begins and ends. Figure 11.1 illustrates conceptually how these systems are organized within a utility organization.

Access lists within each access point are meant to control the flow of information and prevent unauthorized data from entering and leaving each of these networks. Many times, these entities find themselves needing to use resources in other networks. As a result, ports and services are opened between them so that such transactions can occur. For example, the SCADA system may send information to a historian system that may reside on one of the other networks. This examination of data passing

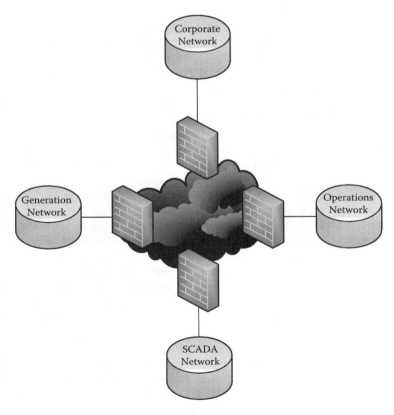

FIGURE 11.1   Access points.

through the firewall from within a network to another network outside the firewall is known as *egress filtering,* while the inspection of network traffic allowing the traffic into its network is called *ingress filtering.* In both instances, the network that is transmitting the information requires that a port, service, and whether or not the traffic is allowed or not, to be specified. This rule definition must occur on both sets of firewalls, and this is essentially how access is controlled from one network to another.

Access points that are firewalls usually use something called "stateful inspection." Stateful inspection is the concept where egress access must allow ingress access that is monitored by session, meaning that the entire connection stream is monitored to ensure that the state is maintained. A connection to a device outbound is then allowed back inbound, as the inbound connection is authorized by the fact that an outbound connection was allowed. Different ports may be necessary for the return traffic because of the manner in which it must be transmitted back. So one may attempt to download a file on port 80, but the download comes back on port 10000 (or any number of ports). The download may also come back on the same port. The point is that the firewall monitors the initial connection, ensuring that it is the connection that the host wishes to make and automatically allows return traffic to that host on any number of ports, depending on how the remote host is configured. In this context, the firewall is said to "maintain state" or remember the outbound connection and allows in the return traffic if it is associated with the outbound traffic. Firewalls are generally configured to allow more traffic going outbound, but to restrict inbound traffic unless the stateful inspection process is enabled. If the stateful inspection process is in use, then inbound traffic will not be as restricted when an outbound connection is made.

Compromises of systems within networks usually occur as a result of user behavior. In a corporate environment, users may download software from a website that is infected with malicious code of some sort. An example would be where a user downloads software from a "Warez" site to gain access to software without a valid license. The software package is then installed on the user's machine and the infected file causes the machine to become infected. The file basically installs another set of files on the user's system so that the attacker can maintain some ability to make the user system perform operations at the attacker's command. The malicious code can do any number of things to the machine, depending on what it was programmed to do. This malicious code can convert the machine to an automated software agent, also known as a *bot,* which is then used to do the bidding of some other machine on the Internet somewhere. Because the malicious code is now on a user's machine, it can access the Internet (if Internet access is allowed) using whatever it has been programmed to do. The programming might tell the code to access a certain IP address on the Internet, so that when the machine is on and connected to the Internet, it automatically looks for the remote host. Many times these programs will beacon to the remote host that they are ready to communicate. When the attacker gets the signal, they receive a transmission from the user's machine on the remote host and send usually whatever they want back to the user's machine from the remote host; an example would be where the remote host is contacted and sends a command back to the host, which allows the remote host to communicate with the user's machine. The result is that the remote host now maintains the same rights on the user machine as the user.

If the user maintains administrative rights on the machine, meaning that he can configure the operating system and install new software, this will give the remote host the same capability. Through this process, the attacker might upload additional software that may be used to scan other devices in the network that the user's machine sits on. This technique is known as *footprinting,* where the attacker is determining where he is on the user's network. The attacker might first obtain the IP address of the user's machine, including the gateway, which is the router or switch that the user's machine connects to in order to get somewhere else on the network. With this information the attacker will likely be able to determine the IP range that the user's machine is part of and then upload scanning software that is used to scan the IP range. This scanning technique is used to determine which other hosts on the user's network maintain IP addresses. So in what is called a "Class C" address space, there might be 256 possible IP addresses. The scan will tell the attacker how many of those addresses are actually in use. When the attacker determines which addresses are in use, he might send out an additional scan that tells him what the operating system is on each of the hosts that were found to be "alive." The attacker would then move on to try to determine which ports and services are open on the remote hosts, providing him with information that will help him identify the operating system platform, the version, the patch level, and other features. If a certain patch is missing, for example, the attacker might upload what is known as the exploit for a system with the missing patch.

Launching the exploit against the vulnerable system allows the attacker to take advantage of the weakness in the system. The exploit may be used to establish backdoor access and then the attacker can set up another backdoor that he can use without launching the exploit again. This new exploit might not be an exploit at all, but rather an application that simply allows remote access. This technique may allow the attacker to compromise multiple machines. A determined attacker may be in the network looking for something specific. In a utility, the attacker may be looking for systems that maintain connectivity into more sensitive networks such as SCADA, operations, or generation. He may continue executing this process until he has access to the desired resource. Once he has such access, he might just leave other malicious software on it and come back and use it at a later date. He might sell the access that he has created, or he might consider using it later for another purpose. The point is that a good hacker can maintain access to many machines throughout multiple companies and might attempt to profit from it on the black market. The primary threat in such a case is to whom the attacker sells this access. He might sell access like this to a terrorist group, rogue nation, or criminal enterprise that may use it to cause harm, while the motivation of the original attacker is simply to make money. And this is where the threat can have a significant impact on Smart Grid. It should be pointed out that this illustration represents an advanced threat and that the example illustrated a process that was executed against a network with no security controls.

The question then becomes: who would want to buy access to a utility's SCADA system and for what purpose? In the example, the malicious code that was downloaded by the unsuspecting user was just one of the threats. The risk of the threat could have been reduced had malicious code protection been loaded on the machine. In some cases these attackers are smart enough to write malicious code that is not detected by signatures that a typical malicious code protection can identify. When you factor in the fact that there are millions of applications that can be potentially installed on a user machine, then you also realize that there are millions of possible variations of exploits that can be written. But the primary defense in this case would be to not allow the user to access and download the malicious software in the first place. This requires that you control what users can access and when they access them, and what they are allowed and not allowed to download. If the user was denied the ability to download the initial malicious payload, then he likely would have never been able to install it and the exploitation process would likely have never begun. Similarly, this can be extended to the use of application whitelisting, where only approved software can be executed on the machine.

Various incidents have occurred where attackers have sent out millions of e-mails with links to websites that were then clicked on by unsuspecting users. Some users downloaded the malicious package and some users did not. Sending it to millions of users increases the likelihood that some percentage of the recipients would at least attempt to download the malicious package. Proxy servers are used to control the type of content that users are allowed to access on the Internet. This process, however, requires that every malicious site in the world be cataloged and blocked. Proxy server vendors are fairly good at identifying and cataloging these sites, but again there are millions of Web addresses out there that represent millions of possibilities, so all of them cannot possibly be blocked by the proxy server. The point is that there is a chance that the initial attack could occur without it being identified by malicious code protection processes. This should illustrate that the attack is possible on a system connected to the Internet and that provides the context that demonstrates that sophisticated attackers have a number of tools and techniques at their disposal.

Implementing the standard malicious code prevention tools, however, mitigates the risk of the common threats that exist. A standard malicious code prevention tool will prevent most infections from occurring because the vendor that makes them has likely identified most of them and provided signatures to identify and stop them from doing harm, even if the user's machine is compromised. The problem continues when these tools fail to identify the threat, thus allowing the attack to continue. As a result, the utility may choose to implement a second line of defense, which might be an intrusion detection system (IDS).

The IDS monitors network activity, looking for potential malicious code being sent and received by machines. When this malicious activity is detected, the IDS alerts whomever it is programmed to, so that a response can be issued. The identification of such an event might be correlated with the current signature base enabled on the malicious code protection tool. For example, if the IDS is programmed to identify traffic on ports that are not commonly used by the business, and it sees one of these ports used along with a payload that is associated with a threat, it might generate an alert identifying that a threat exists on the network. Many times, IDS devices will generate a great deal of alarms, so tuning of that device to eliminate what is commonly referred to as *false positives* is important so that only real potential threats will be identified. False positives are identified because some traffic that looks malicious may just represent activity that is normal for the systems being monitored. For this reason, the implementation of such a tool must first be baselined against the network that it is monitoring. The key in such a scenario is when it detects something that may represent a potential threat and is not representative of normal behavior, that information from the malicious code protection tool is correlated with the IDS threat identification process. If the malicious code protection tool identifies that some hosts have been infected with a threat and the IDS indicates that the threat exists on the network, then there is a higher probability that the threat is a real threat.

An attacker may attempt to circumvent these controls by attempting to program his exploits to seem like normal network activity. This may require the utility to implement additional controls to identify threats that the malicious code protection tool and the IDS fail to identify. Another such tool would be in the management of security patches on the hosts, including patches for operating systems and applications. If the security patch management system can report on where the current security patch level is, which is representative of whether or not the most current security patches have been applied to the operating systems and applications, then the risk of infection on a system that is not currently patched can at least be identified. It becomes even more complex when you begin understanding the concept of patches. Security patches are released by vendors to resolve flaws in the product code for which there is evidence that some malicious code has been developed and used to exploit the flaw. The problem is that not all forms of malicious code are patched. The bigger vendors monitor this and attempt to release patches to address these issues, but it is literally impossible for a patch to be developed before the threat is developed. And for smaller vendors that do not have the staff to keep up with this, patches might never be released unless they become major issues. The bottom line is that there may not be patches for all the vulnerabilities that might be available for a current platform. This may require that additional controls be implemented in order to help in managing the risk of not maintaining the current patch level.

The next control is in the limitation that is placed on access. Good security practices dictate that access into and out of an environment be restricted to only the port and service that is necessary in order to accomplish the mission of the host. As hosts will have access to other hosts, restrictions may be placed around critical hosts to prevent some communications between certain systems. For example, if a user machine needs to communicate with a directory server so that the machine can check the directory to perform identification and authentication, it may do so on specific ports. The switch that brokers access from the user machine to the directory server may be isolated with access lists. Access lists will prevent the user machine from communicating with the directory server on all ports except for the port required to validate credentials. This represents a security control because allowing all communications from a user machine to a directory server provides a sophisticated attacker the chance to try exploits using different ports and protocols. If this access is restricted to specific access, then the attacker may be forced to attempt her exploit on the port that is needed for business purposes. The IDS can then be programmed to look for payloads that do not represent what would be expected between the two machines. This concept of controlling access within the network by switches is usually called *segmentation*.

Segmentation is usually used to segment user machines from servers and applications. As only certain activity is needed to communicate with servers from user machines, the switch can be programmed to only allow this type of traffic. While this control does not necessarily prevent anything from happening, it does make the use of the network more precise, which further limits an attacker's possibilities when attempting to compromise machines and hosts on the network. The more controls a company can apply, the lower the risk to their assets become.

11.2.1 Malicious Threats in Control Systems

The primary malicious threats to control systems are networks within the utility that allow for day-to-day interactions with the Internet. This is because these environments are exposed to cyber-risk on a daily basis, which increases their chances of compromise by various threats. The previous section centered on how a user on a user machine would get infected with malicious code and how an attacker would use that infection to gain additional access to corporate resources. Using such access to compromise a control system represents a unique challenge because the systems are isolated from the network where users reside. This is better represented by comparing two rooms right next to each other. One room is where your critical food supply resides and the other room has a few rats in it. The house in which the two rooms reside is surrounded by rats. So the question becomes: how long will it take for the rats to enter the house? And if they enter the house into the first room, how long until they get to the room with the food in it?

These questions illustrate the challenge that utilities face. Although access is controlled by a firewall that appears to be the only way into a control system network, there are a number of other potential access points that could represent possibilities. For example, these user machines might sit in the same room as the control system network but are connected to separate switches that allow for separate access. The first issue in this scenario is that only the network technician who set up the connection knows whether or not there is really proper segmentation, and there is the potential that the network technician has chosen to connect everything to the same switch and perform traffic isolation within the switch itself. Assuming that everything has been set up properly and the appropriate verification has taken place, there is still the risk that users need to get data gathered from external resources to internal systems. They may attempt to do this by downloading their files on a USB (universal serial bus) key and then pulling it out of their machine and placing it into the control system machine in order to copy the data across machines.

The USB key that was used might potentially be the USB key that is used by the user to maintain a wide range of information, from music to movies to files. The point is that there may not be any security control in place to prevent the user from doing this, and the malicious file may reside dormant on the USB key until it is used. Something else to consider is that these control system networks are rarely infected by malicious threats; therefore, the probability of a compromise is seen as being low. As a result, patches, updates, and other operational security components are implemented at a slower pace. This slower pace means that if the threat were ever introduced to the control systems environment, it is likely that it will represent a serious problem with the control system.

To mitigate these risks, utilities need to identify what the potential threat vectors are that are associated with control systems. Threat vectors are effectively access points or opportunity points that can be used in order to allow the malicious code to be introduced into the control system. The following examples represent potential "threat vectors":

- USB drives
- Access points
- Vendor access points
- Dial in access point
- Remote access points
- Physical access points

To mitigate the risks associated with compromise, utilities should consider addressing how they will prevent these threat vectors from being used. The second question that should likely be asked is: if these threat vectors are necessary for business use, how will they detect that a cyber security event has occurred using one or more of these vectors as the attack launch point? The utility will want the ability to identify that something has occurred and develop a strategy to deal with it. The basic question that should always be asked is: how could a threat be introduced into the control system? This question should be continuously asked, and there should always be a response to the question, that is, the control that detects or prevents the event from occurring.

The danger in Stuxnet, a highly sophisticated, multi-part attack on a specific control system,[1] and threats like it is that they are sophisticated threats designed to automatically exploit the system that it was designed to compromise. Your typical malicious threat may open a backdoor, allow attackers to remotely access the system, look for opportunities to attack other systems, and eventually create the desired incident. Stuxnet and threats that are considered *weaponized* can basically do everything that a remote hacker can do without having to remotely access the system that will be compromised. Furthermore, these threats are programmed to attempt a limited number of exploits and perform the attack over an extended period of time, so that it is not detected by conventional security controls. The true danger in a weaponized threat is that it is programmed by someone to create a specific impact. In a control system, that impact might be to shut down the entire system. If the attacker programming the threat knew what steps he needed to take in order to shut down the system, those steps could be programmed into the threat. An attacker might wish to simply force information to change in the system, which eventually causes operator error. There are a number of possibilities; the question is: how would they get such a tool into a control system environment?

The sophistication of Stuxnet and other weaponized threats indicates that the cost to produce something similar will be significant. Moreover, there is likely a cost associated with getting the threat into the control system environment. As utilities are getting better and better at securing against most cyber-threats through the implementation of more and more sophisticated security controls, attackers who have a desire to compromise a control system might start looking toward bribery. One mechanism that attackers could use is to identify their target on the Internet and look for human targets within the targeted organization that might have the appropriate access. Accessing social media sites related to professional experience within any industry would likely lead a threat to identify who within a company is a "control system operator," for example. They might approach that control system operator with a bribe request, a blackmail threat, or they might simply resort to physical intimidation to coerce the operator to insert a USB key on their control system. The point is that there are a number of physical attacks that could be blended with cyber-attacks in order to create the undesired impact.

While the risk of an event such as Stuxnet on a U.S. utility company is low,[2] given the controls in place, the fact that it could happen should cause pause. The capability exists. All that appears to be lacking is the motivation and the opportunity. As the motivation and opportunity increase, the capability will be more sought after and become more profitable. The capability's profitability is then directly related to motivation and opportunity, which is illustrated by demand. There is a limited capability for executing such an attack that represents the supply, and there is a demand that is represented by motivation and opportunity. As the demand for this capability increases, so will the risk associated with a possible attack. As a result, it is the demand that must be managed; keeping this demand low will result in reduced risk.

As Smart Grid becomes reality, the capability needed to attack the grid will increase, which will make it available to more people. The capability increases because of the convergence of the communications medium necessary to support a smarter grid. The implementation of IT applications designed to automate processes and procedures, which ultimately make the grid smarter, makes for more targets of opportunity. Because of the costs involved in maintaining separate networks, we will

likely see network convergence between corporate, operations, generation, and the control center networks. While these networks will never
operate as one, additional ports and services will be opened in the access
points that support them so that communications will be enabled across
multiple systems. The control center system, for example, will likely need
to leverage remote terminal units that will be represented by applications
within other networks. The legacy RTU (remote terminal unit) architecture relies on point-to-point communications between RTUs and field
assets. As these RTUs morph into applications and begin using IP layer-
based communications protocols, it is likely that the control center will
need to connect to them in a manner that is different from the traditional
method. This is because switches, routers, and other network communications devices will likely be used to support the new infrastructure. The
key to containing malicious threats is then in the deployment of security
controls that are advanced enough to detect and potentially prevent the
systems from being compromised.

11.3 NONMALICIOUS THREATS

Nonmalicious threats are those that could accidentally present a problem to control systems because of mis-operations. These threats can be
coupled with malicious threats due to potential integrity issues that may
be caused by malicious threats. If a malicious threat alters information
that operators are reliant upon to make critical decisions, resulting in
an incorrect decision being made, then there could be an adverse effect
on control systems. As discussed previously, an operator may choose to
open a breaker because of the information that he sees on his display.
This represents a data integrity issue that will likely become more prevalent as smart applications enable the Smart Grid.

Mistakes are always made when supporting anything. The key lies
in reducing the number of mistakes that can occur through the development of processes and procedures that, when followed, facilitate
reliability. Even with processes and procedures, variances to them can
still occur. It is those variances to the processes and procedures that
represent the largest opportunity to identify a nonmalicious threat to
a control system. If process dictates, for example, that operators check
multiple systems to ensure the information that they are using to make
a control decision, any inconsistencies in the check will likely result in
operators not performing the intended action. NERC (North American
Electric Reliability Corporation) has developed a set of reliability standards meant to standardize processes and procedures associated with
grid management. While the totality of these standards does not dictate
how a utility will manage the grid, it does require that processes be

developed to ensure compliance. Compliance assurance represents that
the utilities are required to meet the standards by following procedures
that will ensure a consistent approach to the management of the grid.

The emergency operations planning (EOP) standard, for example,
requires that transmission operators "develop, maintain, and implement a set of plans for load shedding."[3] These plans are designed to
indicate how and when a utility will shed load that might be required
given some event. The plans should illustrate how an operator identifies that load shedding is necessary and what the operator should do in
order to actually shed load. If load is shed and the process is not followed, then the utility will likely be found out of compliance with the
NERC EOP standard. At the same time, however, the utility should see
this as a nonmalicious cyber security event. This is because the operator would have done something out of variance with the process that
may have led to an incident. For example, if a plant operator sees that
one of the utility's generation units has failed, the process that he must
follow might be to contact a technician to verify that the event actually
occurred. If the operator simply sheds load because his system gave him
an indication that he had to, and the system was incorrect or a malicious threat was intentionally reprogrammed to the system to report the
need to shed load, then an incident might occur. The incident might be
that load was shed for a set of customers because one of the generation
units went offline, when in fact load was shed and the generation unit
remained operational. If the operator were to follow the required process, then it is possible that he would have learned that his system provided him with inaccurate information. This example illustrates how
a variance in the process could result in an incident. To detect such an
incident before it occurs, the system might be programmed to notify
multiple people if an event is detected and require the personnel notified
to contact each other.

Smart Grid will likely result in the automation of such a process,
wherein a system is notified of a failure and it acts as a result of that notification. As a cyber security control, a utility might require that the system
notify and request verification of the event from multiple systems before
it is allowed to react to the potential problem. Of course, timing comes
into play as the notification and responses must occur in the amount of
time needed to respond to the event. If the event must be responded to
within 15 minutes, for example, a utility might need to build in a process
that allows for the system to make the decision in an adequate amount of
time. The point is that processes will need to be reviewed for opportunities to insert additional steps to verify that the actual process is being
followed. Where it is found that some process is not followed, the system
should alarm staff that something may be potentially occurring so that
the follow-through on that process can be monitored.

A security operations center that can see these alerts and alarms will likely need to be established and correlated to other security event monitoring systems so that incidents can be appropriately identified. It is important for a utility to be able to identify a nonmalicious incident because it may be an indication of a malicious incident occurring.

11.4 INCIDENT IDENTIFICATION

Identification of an incident, whether malicious or nonmalicious, needs to occur in order to support the reliability of the grid. Generally, malicious events are managed and monitored by cyber security staff while nonmalicious events are managed and monitored by control center operators. Cross-training of the two departments and potentially convergence might need to occur in order to gain a better grasp of the potential that something might occur. Utilities might start considering the convergence of the control center with the security operations center, for example. As the threat identification capabilities of cyber security professionals within utilities mature, control center operators should likely be trained on how to address these alerts. For example, if control center operators are alerted when bandwidth suddenly spikes on a communications network, indicating that a denial-of-service attack has occurred, they will be have a better understanding of why something else in the system has failed as a result.

Security operations must work on developing strategies associated with the identification of cyber security threats and socialize those strategies to the operators. Operators will likely have to illustrate their processes and procedures to cyber security professionals so that they can be retrofitted to detect potential incidents that may be nonmalicious in nature. If both parties can develop a strategy for dealing with malicious and nonmalicious threats, a capability can likely be developed that will augment the identification of an incident. The earlier an incident is identified, the easier it will be to deal with. A joint strategy will help reduce the overall risk to Smart Grid through the development of approaches to identify risk. These approaches can then be used to develop processes to manage the security of operational processes.

11.5 INCIDENT CONTAINMENT

When an incident is identified, it must be contained in order to prevent it from further proliferation. Whether the incident was caused by a malicious or nonmalicious threat, it must be contained to ensure that it does not continue to occur. When there is an availability issue, the containment of the incident is straightforward: stop what is occurring so that the system can continue to operate. In addition to this, a root cause analysis should be executed in order to understand what the threat did and why. The what and the why will be very important when it comes to ensuring that the incident cannot occur again. Depending on the event, the root cause analysis would work back from what has occurred. For example, if the incident resulted in the shutdown of the control system, then the first step in a root cause analysis is to determine why the system was shut down.

A denial-of-service attack against the primary switch in a control system network would likely result in the systems within the control system not being able to communicate. This is because they must send and receive their messages through the switch, which is the component in the architecture that effectively enables IP-based communications. If the switch is unavailable, then the components of the control system will likely not be able to communicate with one another. Some of these components may depend on other components in order to function. If the system that displays the line diagram of the grid being monitored by operators in the control center is not able to gather information from the applications it is connected to, then the operators will not be able to tell what is going on. This may be one of the effects that could be observed if a denial-of-service attack was launched against the switch. Without this information, operators will likely have no visibility into the status of the grid that they are required to monitor. While this in and of itself would constitute an incident, cyber security investigators would then ask questions to determine why the systems are not displaying the information that is usually displayed.

There could be a number of reasons why the display to the operators does not function as normal. The job of the investigator is to first determine why it is not working. With a list of possible contributing factors, an investigator might be able to quickly determine that there is an issue with the switch that connects all the devices within the control system. For instance, he might try to "ping" a system from another system and find that there is no response from the system to which the ping was sent. This will likely result in the investigator attempting to run a *trace route* to determine where communication breaks down. The administrator might find that he is unable to reach the default gateway, which represents the switch or router that is brokering communications. This will likely lead the investigator to investigate the status of the switch. He may find that it is not working properly and reboot it to correct the problem. In the case of a denial-of-service attack, he will likely find that the reboot fixed the problem—until the attack starts again. The fact that the switch functionality cannot be restored with a simple reboot would likely imply that there is a wider problem. The investigator might then decide to identify what is specifically communicating with the switch and find that there are a host of devices within the network sending packets to it. To contain the incident, the investigator might simply remove all the physical connections to the switch or gateway. This would assume that the attack was a denial-of-service attack, that the bots in the attack are inside the network and programmed to send numerous packets to the default gateway. This example is then meant to illustrate a potential use case in order to provide context when running through a root cause analysis.

Removing the connections to the switch will have the effect of stopping the attack if the switch were being bombarded with packets from other devices in the network. This strategy will allow the investigator to reprogram the switch, changing the IP address of the gateway so that it is no longer a target. This strategy would effectively contain the problem, but there would still be the issue of the bots in the network. Each system within the network would have to be analyzed to determine what it was doing from a communications perspective. Where malicious communications were discovered using a *sniffer*, the investigator could then shut down those systems in an effort to stop the attack. The investigator might conduct research on the problem or apply forensic techniques to the system to determine why it is sending such communications. The investigator might find in this case that malicious code was installed on the system, which is what caused the problem in the first place. The malicious files might be removed from the system, or the system might be rebuilt in order to stop the incident from occurring once it is placed back on the network.

As the causes of the incident are traced back to the origin, the root cause of the entire incident is eventually uncovered. The root cause of the incident may end up being where an employee used an infected USB key on several of the control system workstations. This might be detected through interviews and analysis of the malicious code that may have been identified on the workstations themselves (assuming the workstations were implicated as sending the denial-of-service attacks to the switch). The root cause of the entire incident might then be determined to be a combination of not having malicious code protection on the workstations and not controlling USB keys that can be used on the workstations. Assuming that the root cause was properly identified, the investigation can proceed to eradication of the threat, so that it can never be repeated again.

11.6 INCIDENT ERADICATION

Incident eradication refers to the removal of the threat by eliminating the root cause of what permitted the attack in the first place. Going back to the example, if the root cause was that a user used an infected USB key on multiple workstations that did not have malicious code protection, then a mitigation plan might be to remove the ability to plug a USB key into a workstation and implement malicious code protection. Together, these two mitigation strategies would ensure that the same threat could never be repeated in the system. Additionally, further protection was afforded to the system because USB keys could no longer be used as a threat vector and the malicious code protection would help to prevent any future compromise by similar threats. The point in this case is that the attack could never be repeated and lessons would have been learned from the incident. Reliability would thus improve as a result. Of course, if USB keys were being used for a legitimate business purpose, such as moving data between isolated networks, then an alternative must be devised that does not shift the same problem to another method of moving data between segregated devices. For example, requiring individuals to use compact discs rather than USB keys may not reduce the risk much, if at all. However, where USB keys are not used for a legitimate business purpose, the easiest solution may be to disable or remove the USB interface on the computer systems.

11.7 CYBER-DISASTER

A cyber security disaster in a Smart Grid would have a significant impact on society. This is because the disaster would potentially affect the core of society, which is the use of electricity. Without electricity, our way of life would basically cease to exist because almost everything around us is based on the use of energy. We use electricity to power our homes and offices, filter our water, and process our foods. Without electricity, people would likely starve, and we would remain in a constant state of emergency. The federal government would even have a difficult time maintaining order because it would have issues managing the disaster without electricity. This all of course assumes the worst, in that no electricity is available for anyone to use.

Smart Grid, as with the current electric system, is distributed, which reduces the likelihood that something as significant as a total outage could actually occur. The eastern and western interconnections are separate from each other, meaning that one cannot affect the other. In theory, if you had a cascading blackout in the eastern United States, it would likely not affect the western United States. By contrast, the concept of the balancing authorities makes it such that faults will be isolated to a certain area, as utilities have the capability to island potential problems. To create electrical Armageddon, an attack would have to be coordinated to hit almost every balancing authority in the United States at the same time. And the attack would have to defeat all of the cyber security controls that exist in multiple utilities in the same time frame.

Electricity moves in a continuous cycle that, when interrupted, simply moves in another direction. Because there is a lot of monitoring associated with the grid, utilities will likely respond to an incident during the initial event and would likely not allow the entire system or interconnection to shut down. What we could see, however, is a disruption of service to multiple generation resources, or perhaps a meter outage that affects a localized part of a city. These localized or regional power incidents could cause the perception of an unstable system and create turmoil for the industry. While utilities do focus on resiliency and the ability to recover, they will have to understand the mentality of the general population and manage that risk accordingly. Having an appropriate response and recovery strategy, with a focus on limiting the amount of downtime, will likely go a long way if a cyber security disaster ever occurred on the grid.

As the grid is more likely to experience incidents in different systems that may cause outage problems and issues for the general public, the responses should focus on restoring capabilities. Cyber security professionals will need to understand that they must augment the utility's contingency plan in order be effective in and during an incident. Utilities have been good at planning for contingencies for a long while. As their focus is reliability, they will usually maintain multiple redundant systems. For example, if load is lost within a balancing authority, the utility knows where to buy additional load from and will be able to gather that almost immediately. This has to do more with the dams, which can create almost instantaneous generation simply by opening a valve and releasing water into to it. As discussed in previous chapters, the manner that hydroelectric generation works allows for near-instantaneous power to be available in the event of a massive load-shedding event. As a result of this capability, hydroelectric generation units allow for the availability of a limited amount of generation in the event that load is shed and must be made available. This all depends on the lake levels that the dam supports, as the water level must be above the turbine so that the valve can be opened and water can be drained into it, thus allowing the turbine to spin.

NERC BAL-002 requires utilities to maintain contingency reserves, or to know where reserves can be purchased in the event of loss of load. A contingency reserve is required for the reliable operation of the interconnected power system.[4] Adequate generating capacity must be available at all times to maintain scheduled frequency and avoid loss of firm load following transmission or generation contingencies.[5] This generating capacity is necessary to replace generating capacity and energy lost due to forced outages of generation or transmission equipment.[6] These organizations are continuously preparing for a disaster and know how to restore energy in the event of a loss of load condition. When load is lost in one area of the grid, areas where generation remains available will allow for restoration of the load, even with a loss in generation. Balancing authorities are required to maintain a minimum contingency reserve, which is the sum of the amount of reserve equal to the loss of the most severe single contingency, or the amount of reserve equal to the sum of 3 percent of the load (generation minus station service minus Net Actual Interchange) and 3 percent of net generation (generation minus station service).[7] This is one of the many standards written into the NERC BAL standard, which is meant to ensure that appropriate generation is always available.

11.7.1 Load-Shedding Events

A cyber-disaster in Smart Grid is really a situation where load is lost and enough generation is not available to compensate for that loss. Load shedding or loss of power can result from any number of areas surrounding the Smart Grid. Hacking into and disabling enough smart meters could result in a major load-shedding event. If an attacker were able to gain control of a smart metering infrastructure and issue disconnect commands to enough households, a major load shedding event could be triggered. This disconnect could trigger a great deal of loss of load that would, in turn, force the utility to respond by reestablishing power to the disconnected meters. With the introduction of automated load control, it is possible that the distribution system could impact the smart meter infrastructure in such a manner. This type of disaster would likely be limited to an affected area and would probably not be enough to bring down the entire bulk electric system for one of the interconnections.

A disruption in service on the transmission system, however, could have a greater impact. One risk might be where a series of 500-kilovolt lines were taken out of service by a threat that disabled systems within a substation. As these 500-kilovolt lines represent a significant amount of power to multiple service areas, the result might be a major load-shedding event. This represents the rationale for the NERC standards, as the whole point of these standards is to prevent an incident that would affect the bulk electric system. Maintenance, contingency reserve, cyber security, and other standards are meant to prevent such an incident, as it might have a much larger impact on the greater electric grid. Even in such a case, the load-shedding event would likely be isolated to one interconnection. The real danger associated with a cyber security event to the bulk electric system would be where an attacker executes a coordinated attack that triggers a major load-shedding event in multiple interconnections and results in multiple cascading blackouts all at one time. This would effectively be the Smart Grid Armageddon, as it would really require all utilities to coordinate in order to restore power. Even in such an event, however, the power system would likely be restored; it is just a matter of when. It is possible that such an event would disrupt electric service for many for an extended period of time. This is likely because of the level of coordination and work that would go into system restoration.

11.7.2 Cyber-Disaster Response

Utilities maintain the capability to restore their systems in the event of a load-shedding event. NERC regional entities, for example, are required to maintain a system restoration plan, which translates into emergency operating procedures (EOPs). The system restoration plan requires utilities to maintain what is known as a *blackstart capability*. The blackstart capability requires utilities to understand how they would restore power if all power were lost. This represents the single worst event that a utility can experience. There is also a regional blackstart plan, which represents how the different NERC regions would restore power based on the plans developed by the utilities within each region. For example, if all power was lost, some power would be needed in order to restore larger generation so that full power can eventually be restored. Hydroelectric generation might be used to start up a line that feeds directly into a coal plant that maintains the ability to produce more generation. Once the line from the hydroelectric plant to the coal plant is started, it creates the ability to energize other lines and paths from the coal plant to other sources of generation. The point is that the grid can be systematically restarted through the blackstart process. And this blackstart process is meant to be the worst-case scenario, where all electricity on the grid was lost.

A cyber security incident would likely also trigger a cyber security incident response plan. Going back to the examples that were provided earlier in this chapter, an incident response capability would have to be initiated so that the reason for the blackout can be established. If the root cause of the event cannot be determined, then it is possible that the event could continue occurring. For this reason, a forensics investigation might be necessary to make a determination of why the event occurred so that future events can be prevented. If a decision is made to restart operations before the root cause is known, the event might occur again in a short period after the first event. Because of the potential for this problem, incident response and recovery planning become very important. The more planning in anticipation of such an event, the better prepared the country will be to respond.

The planning of such an event might be represented by identifying levels of impact. If one power plant is taken offline, it might represent a low impact to the grid. If two or more plants go offline as a result of a cyber-attack, a moderate impact level might be assigned. The same would be true in the management of smart meters, distribution, and the transmission infrastructure. NERC currently defines this in CIP standard 002, which requires entities to execute a Risk Based Assessment Methodology (RBAM), in an effort to identify critical assets, which may represent a high impact to the bulk electric system should a cyber security incident occur.

The RBAM is meant to illustrate the worst-case scenario and thus require cyber security protection around those systems. There is, however, speculation that impacts on assets that might have a limited impact on the bulk electric system could translate into a serious problem if they are successfully targeted and exploited. This represents a gap in the development of an impact identification process for the bulk electric system. Identifying impacts on the lower impact assets and aligning them to the larger impact assets would represent a risk management process that could be used to indicate the severity level of an event. If a smart metering infrastructure, for example, is disrupted because of a cyber-attack, the utility might respond by representing the attack as a yellow level alert. This yellow level might represent an impact to the localized power system but not the bulk electric system. The key lies in knowing what an impact means should one occur.

11.7.3 Cyber-Disaster Recovery

Utilities have for years been preparing and testing their ability to recover from a power disturbance. In fact, the mission of most utilities that support grid assets is in the delivery of reliable power. The more reliable a utility's power assets, the more the mission is supported and sustained. The effects of cyber security on the reliable operation of energy infrastructure are not well known or understood. Historically, these systems have been separated from cyberspace because there has really been no need to connect to it. As information technology starts to play an increasing role as a result of Smart Grid, the need to leverage cyberspace will become increasingly important. As a result of this newfound importance, Smart Grid will become more and more exposed to cyber security threats. Furthermore, as the benefits of Smart Grid grow, so will the motivation to threaten the infrastructure. With this growing threat will come the need to understand how to recover systems when or if a cyber security incident ever occurs.

Cyber security and IT professionals will need to understand what can go wrong on the power system and understand the recovery options that are available for the assets that support it. In most cases, the recovery of power system cyber-assets will differ from typical IT assets. They may be required to support the recovery of these assets through an approach that is integrated with the power system priorities. For example, if a smart RTU fails and the strategy to correct the failure is to replace it, how long will it take to get a new server and install the software to get the RTU back in service? The recovery priorities for each system that supports the various aspects of the power system must then be defined as well as or better than the current definitions available in the IT world. Impact priorities, for example, must be known and will likely be much more aggressive than they currently are for traditional IT assets. The fact that these systems are distributed to different areas of a utility, in the field, in a plant, and in a control house must be considered. IT organizations will have to understand how they can support this distributed architecture and respond in a time frame considered reasonable for "real-time systems."

ENDNOTES

1. For a more detailed description, see Chapter 7, Note 42.
2. In this case, the risk is that the sophisticated exploit is targeting a particular system run by that specific utility. As it turns out, a large number of utilities had computer systems infected by Stuxnet; but because those systems did not employ a very specific type of control system hardware arranged in a particular sequence, the impact was relatively benign.
3. For more information about NERC Emergency Operating Planning, see Standard EOP-001-0 — Emergency Operations Planning, at http://www.nerc.com/files/EOP-001-0.pdf, April 1, 2005.
4. For more information about NERC Disturbance Control Performance, see Standard BAL-002-1 — Disturbance Control Performance, at http://www.nerc.com/files/BAL-002-WECC-1.pdf, August 5, 2010.
5. For more information about NERC Disturbance Control Performance, see Standard BAL-002-1 — Disturbance Control Performanc, at http://www.nerc.com/files/BAL-002-WECC-1.pdf, August 5, 2010.
6. For more information about NERC Disturbance Control Performance, see Standard BAL-002-1 — Disturbance Control Performance, at http://www.nerc.com/files/BAL-002-WECC-1.pdf, August 5, 2010.
7. For more information about NERC Disturbance Control Performance, see Standard BAL-002-1 — Disturbance Control Performance, at http://www.nerc.com/files/BAL-002-WECC-1.pdf, August 5, 2010.

CHAPTER 12

Crystal Ball Time
Will We Have a Secure Grid and What Will It Take?

12.1 INTRODUCTION

Smart Grid represents the future of energy production and delivery. Customers will be able to interact with the grid, making it a larger part of everyday life. Smart Grid will enable consumers to take advantage of new technologies that have made life easier. As such, Smart Grid will likely introduce some significant benefits to consumers. For example, consumers will likely be able to use their mobile devices to turn on their pool pumps, air conditioners, and other appliances through a remote interface. Consumers will actually be able to access their home energy network remotely, but more importantly, leverage the technology to save on energy costs. Likewise, the utility will see a reduction in cost if all the benefits of Smart Grid are realized. This will allow the utility companies to implement renewable energy, move away from nonrenewable energy production, and save money all at the same time. The challenge in getting to this point directly ties to the initial investments that must be made in order to develop Smart Grid technology, as well as the security challenges that will be associated with the implementation of such technology.

Improvements to the grid will require upgrades at the technology level for the industrial assets that support the overall electric system. As discussed in previous chapters, the technology upgrades applied as a result of the modern grid strategy will have a direct impact on the future of the grid. Changes to smart metering, energy distribution, transmission, and generation will likely represent the most concern from a cyber security perspective. This is because the infrastructure associated with these key elements of the grid represent the most critical part of the modern grid.

12.2 SMART METER SECURITY

If a cyber security incident were to occur within the context of Smart Grid, we would likely see it in the smart metering infrastructure. This infrastructure represents the least impact to the overall bulk electric system but is a risk because of the effects on consumers. The probability of a cyber security incident for smart meters is highest among all the infrastructures because they maintain a great number of assets without any real physical security protection. This means that anyone can approach a smart meter and "do something to it." Furthermore, the networks that support communications within these smart meter networks are not as reliable as advanced communications networks and that represents a potential reliability issue. The security of the communications network is less of a concern when compared to the reliability of the communications network itself. Because the smart meter field communications network relies on radio-based technology, it can be challenged for reliability at times. This is because a radio-based communications network is subjected to interference and other problems that may impact its reliability.

The potential impacts as a result of interference will likely contribute to the first cyber security incident, which will likely be jamming of the smart meter communications network. The impact of such an event will be low, considering that there will be no direct effect on consumers' electricity. This means that electric service would not be disrupted as a result of a jamming attack to the smart meter field network, but it would represent the fact that a threat exists that intends to do harm. The effect of such an attack would most likely affect the utilities' ability to control and gather information in regard to energy usage. This is similar to denial-of-service attacks against IT (information technology) networks that were very prevalent as the Internet gained increasing popularity. As time passes, these attacks will become increasingly prevalent and have a deeper impact on cost, as associated with responding to the problem. These problems will likely lead to utilities pursuing alternative forms of smart meter field communications technologies. The cost of maintaining a 900-megahertz mesh network today is very cost effective. However, the cost will increase significantly if the utility is unable to consistently use it to control and gather information. At some point, the cost of a newer technology that is not prone to a similar attack will outweigh the cost and benefit of maintaining the radio-based communications network. When this occurs, the communications network will begin changing to something more advanced.

12.3 HOME AREA NETWORKS

Home area networks (HANs) will play an increasingly important role in the near future and will likely represent the front line of attack for cyber security threats. Just as people figured out how to break into home routers, they will likely find ways to break into the HANs. It is likely that the management of the HAN will be left to the consumer, which may result in the lack cyber security controls to mitigate risk on the HAN. Similar to the choice consumers have on their 802.11 wireless networks, they will likely have the option to enable encryption and other security controls to protect their home network from threats. Consumers may experience problems in connecting to the devices that they wish to control remotely and those issues may be traced back to the fact that encryption or other cyber security controls are enabled. This may cause the consumer to disable them, which will increase the risk of a compromise in the home.

A compromised HAN represents a relatively low risk, depending on the type of information that may be available. If the HAN facilitates the transmission of any privacy information such as customer account numbers, credit card numbers, or social security numbers, it is possible that a compromise of the HAN could have a higher impact. The likelihood of such an event is solely up to the meter vendors, consumers, and the utilities themselves. Because of such risks, utilities will have to evolve their cyber security programs to encompass communications to consumers themselves. We can expect to then see utilities providing cyber security awareness to their consumers in order to augment the mitigation of cyber security risks in the home. More importantly, utilities should be limiting the amount of sensitive information that is transmitted over these networks. A device serial number and an associated message signed by the device's private key should be sufficient to identify the device and its associated owner.

12.4 HEAD-END AND METER DATA MANAGEMENT

The backhaul from the smart meter collectors to the head-end system will likely evolve to become more secure. Because there is a potential that compromise of the head-end system can result in the disconnection of multiple customers, we may likely see regulation of that system. Regulation would require utilities to further isolate their head-end systems, treating them more like control systems. As the head-end system effectively provides command and control over all smart meters and collectors, disconnects could result in load-shedding events that exceed the thresholds set by NERC (North American Electric Reliability Corporation) today. This means that the possibility that smart meter disconnects could impact the bulk electric system may draw these systems into the scope of NERC CIP (North American Electric Reliability Corporation Critical Infrastructure Protection) or other future regulations. While it is highly unlikely that we will ever see a massive disconnect of smart meters due to the hijacking of a head-end system, the possibility does exist. That possibility is high or low depending on how those systems are secured from threats. A regulatory approach will ensure that some consistency is applied in the security of these assets, but the reality is that the probability cannot be predicted because the compromise of such a system would require a sophisticated attack or the actions of a malicious insider with the appropriate credentials.

The motivation associated with an attack on the head-end system is probably blackmail or military in nature. Can utilities be blackmailed, for example, into providing money in exchange for a threat to not disconnect some or more smart meters? If a threat maintained this capability, what would be the response from the utility? More likely than not, as is speculated in the finance sector, utilities might agree to pay in order to not feel the consequences of such an event. The reputational damage would likely cause great harm to a utility's bottom line because of the response from regulators and the perception of their customer base. From a military perspective, the head-end system might be an easier target to create a localized blackout for terrorist organizations and malicious nation states. It is not that these organizations would ever use this capability, but that they will want the ability to use this capability in case they ever decide that they need to attack the United States. Terrorists might black out a section of a city block so that they can place a bomb in a discrete location without being monitored. A nation-state may just want the capability as part of an attack strategy that might never need to be executed.

There is evidence that the United States, Israel, Russia, and other nations have this capability over other countries. In the book *Cyber War: The Next Threat to National Security and What to Do About It*, written by Richard Clarke and Robert Knake, there are multiple discussions surrounding the use of cyber-tactics for acts of war.[1] There is a discussion about how Israel, in particular, attacked a perceived Syrian nuclear facility. The book makes the accusation that Israel turned off Syrian radar systems before flying in to bomb the facility. Without the radar systems, the Syrian government did not know that the attack was occurring. Cyber security, in this case, was used to turn off defenses so that an attack could be launched.[2] The book discusses similar uses of cyber-tactics during the U.S.-led invasion of Iraq. While there is no proof that these incidents actually occurred in the manner that is described in the book, the concept is easy to understand. And when understanding how cyber security attacks occur, a well-funded sophisticated threat will likely be able to penetrate almost any data network.

The meter data management system, however, probably poses a greater risk because of the information that it processes. As most meter data management systems can process privacy information and connect to multiple systems used to manage power operations, it is also a good target for threats. Because there is a clear financial motivation associated with identity theft, we will likely see threats going after information within the meter data management system itself. Threats might also want to change customer billing information in an effort to harm utilities due to having some issue with them. Environmentalists, for example, may launch such an attack because of their distaste for the use of fossil fuel in the environment. The point is that there are many kinds of threats and multiple targets.

Customer programs might also be impacted as threats may decide that they can have a significant impact on utilities simply by changing the usage programs of consumers. If a consumer has adjusted his use of energy to a specific program, changing that program will likely result in customer service issues for the utility. That is because the utility will end up having to address why the consumer billing program was modified, assuming a threat source was able to hack into and change customer billing programs.

12.5 DISTRIBUTION SYSTEM SECURITY

The future of the distribution system is clearly tied to the use of controls to automate processes and procedures at the distribution level. The biggest risk lies in the automated load control. We will likely see this type of technology used in the near future in order to prevent or reduce the risks of a localized blackout. Automated load control might be used to detect and respond to localized power problems. Automated load control will reduce costs because it will prevent wider blackouts and allow the utility to respond before something happens rather than sending technicians out after the fact. Utilities will at first use this technology in a limited fashion in order to determine how well it works. The benefits will become clear and the risks will become apparent as a result of the use of this technology. Utilities will be forced to implement cyber security controls on the distribution system to protect the use of these newer and advanced programs.

12.6 TRANSMISSION SECURITY AND THE BULK ELECTRIC SYSTEM

NERC CIP, while not perfect, has been the single biggest benefit to the bulk electric system as far as cyber security is concerned. While the NERC CIP standards themselves represent a minimum set of security controls and not complete security of the system for which they are implemented, they have decreased risk to the bulk electric system significantly because they have started the conversation within the utility industry. NERC CIP has made such an impact on utilities that entire organizations are being formed to address the concern. Cyber security professionals are primarily used to address NERC CIP, which has required them to reside within the operations side of utilities. Furthermore, the industry has identified a real gap because most cyber security professionals have no experience with reliability or the power grid and very little experience with industrial control systems that make up most of the automation for the power grid. As a result, the standards themselves have triggered a lot of discussion on the topic and personnel to concentrate on how to secure the power grid and make it more reliable at the same time.

Transmission organizations will likely end up creating cyber security organizations within their business that are tasked with managing overall operational security. Organizations might create positions such as Operations Cyber security Officer, as opposed to the traditional Information Security Officer. The Operations Cyber security Officer might be someone or a group of people dedicated to the cyber security and reliability of operations within a utility. These groups may be independent of corporate security groups so that the focus can be placed on the business itself. They will most likely report to the head of operations, similar to the manner that an Information Security Officer would report to the Chief Information Officer. This organizational change will likely represent the largest impact of cyber security risks to utility companies at the transmission level.

The risks to transmission organizations are clear. A threat could potentially impact the bulk electric system and create a cyber-apocalypse. As the industry understands the threat, many steps have been taken to reduce the risk of a cyber security incident. Utilities are likely to be much more concerned about the strategic impact of such an event, such as over-regulation of the industry and government oversight. If a cyber security incident were to occur in a manner that affected the bulk electric system, the measures taken to prevent future risks of such an event would far outweigh the risks of a cyber-attack.

Because the risk of a cyber-attack is becoming increasingly important and regulation is consistently increasing in scope, utilities will be forced to address the issue. While the probability of an incident that will affect the entire bulk electric system is rather low given the controls currently in place, it is possible for an incident to occur. We may see worms and viruses come out that are directed at the utility industry. These would be things such as Stuxnet and Night Dragon,[3] but developed as worms released on the Internet in an effort to see where they go. We may see the compromise of a transmission system vendor that could result in the compromise of a utility that allows remote access from the vendor.

The advanced applications and use of smart remote terminal units (RTUs) will likely also trigger the identification of threats that could have a direct impact on the reliability of the transmission system. The probability of these threats will not mean that the transmission system is actually compromised, but the fact that they exist will force utilities to do something about them.

12.7 THE DISTRIBUTION SYSTEM AND NERC CIP

As discussed in Chapter 2, NERC CIP does not apply to the distribution system, which comprises much of what we know of today as Smart Grid, and that is no accident or oversight. For decades the local distribution system was the exclusive responsibility of state and localities. In theory, this made sense. Transmission lines often run over long distances and frequently across state lines. Their successful operation affected huge populations, and the generation they delivered needed to be balanced across regions and interconnections. The Federal Energy Regulatory Commission (FERC) was the appropriate authority to see that electricity was not disrupted through these critical pathways. Similarly, the NERC, acting both at the behest of FERC and as a grassroots industry organization, was the appropriate vehicle to develop and oversee the implementation of reliability standards, including those relating to cyber security, for this transmission infrastructure. The idea of applying the same logic to distribution sort of sounded like the federal government being responsible for fixing potholes on a city street. However, as we've seen with the Internet, the concept of local has lost a lot of its meaning, particularly in the realm of cyber security. Where the only harms we were worried about were physical damage, the local claim had merit. Moreover, the resulting harm was relatively local. Now, as a result of sophisticated communications networks that utilities interconnect with each other on, a small cooperative or municipal utility could be attacked from halfway around the world at the same time that a dozen other utilities are attacked through a hole left open by a single utility. And even when a single utility is attacked, the ability to switch on and off power to thousands of homes has the potential to destabilize the bulk electric grid like nothing that had been done previously. While fail-safe mechanisms can be deployed and probably will be successful, the premise for FERC and NERC to regulate cyber security at the transmission level now appears to exist for the distribution and for Smart Grid as a whole.

As of this writing, Smart Grid cyber security standards are very incomplete. There is some high-level guidance provided by NIST (National Institute for Standards and Technology) and others, but the consensus is not there yet. Some utilities accepting federal government grants are in fact subject to cyber security requirements, but those requirements are less than clear and regulatory oversight is limited. Additionally, cyber security laws currently being considered are too numerous to mention and only nibble at the problem from the margins by regulating some utilities serving military bases or giving FERC enhanced powers during an emergency. It would seem that the point of giving FERC and NERC broader cyber security authority would be to prevent those emergencies from happening in the first place. While stopping short of recommending that Congress pass legislation giving FERC and NERC the oversight authority over cyber security issues at the distribution level where the bulk Smart Grid lies, the Government Accountability Office (GAO) clearly lamented the difficult situation that FERC faces when it noted the following:

"The fragmented nature of electricity industry regulation further complicates enforcement of Smart Grid standards and oversight of Smart Grid investments using FERC and other regulators' existing authorities. Oversight responsibility is divided among various regulators at the federal, state, and local level, and FERC's authority is limited to certain parts of the grid, generally the transmission system. As a result, state regulatory bodies and other regulators with authority over the distribution system will play a key role in overseeing the extent to which interoperability and cyber security standards are followed since many Smart Grid upgrades will be installed on the distribution system. Such regulatory fragmentation can make it difficult for individual regulators to develop an industry-wide understanding of whether utilities and manufacturers are following voluntary standards. This is due to the large number of regulators in the industry—FERC, electricity regulators in 50 states and the District of Columbia, and regulators of thousands of cooperative and municipal utilities—and their potentially limited visibility over parts of the grid outside their jurisdiction."[4]

There is no doubt that this is a complex issue, and some officials have suggested that direct regulation of distribution by FERC would be akin to a declaration of war on the sovereignty of some states that take federalism and their prerogatives to regulate very seriously. However, the economic and technological realities make it difficult for public utility commissions and municipal utility boards to claim that the security of a portion of an interconnected grid is a local concern. Energy is bought and sold across regions, and even distribution through the introduction of distributed generation and electric vehicles is blurring lines between distribution, transmission, and generation. Even rates, long the jealously guarded domain of state and local officials, have an interstate flavor as generators frequently compete for the best price across state lines. Finally, the reality is that state and local utility commissions generally do not have the expertise, or the funding to obtain that expertise, to develop and implement cyber security regulations. However, this is not a slight to these commissions who work very hard to understand a wide variety of some very complicated issues and in most cases represent their constituents. In many ways, taking cyber security responsibilities away from public utility commissions would be doing them a favor. Many are being criticized for their lack of cyber security oversight and turning this responsibility over to NERC as the primary oversight body with ultimate enforcement by FERC would allow the commissions to focus on the issues that they know best. So it is with that context in mind that we recommend that Congress pass legislation that grants FERC and NERC the authority to regulate cyber security over all utilities offering electricity to the general public.

In peering into the crystal ball, one would hope that this will happen eventually. However, the United States has frequently been unsuccessful in regulating something at the federal level that was once done at the state level. If it happens at all, the most likely scenario would be some sort of dual regulation wherein state commissions would have the power to enact more restrictive rules over and above what is implemented at the federal level. Additionally, it is likely and advisable that any security controls mandated for distribution be customized to address the unique needs and cost constraints of the distribution network, as well as the unique risks. In many ways, the newer technology brought in through Smart Grid deployments will likely address many of the security requirements simply based on what is built into the technology. However, technical feasibility exceptions based on technical and economic concerns will still be warranted. As many have already done, utilities should begin to think of how to address NERC CIP at the distribution level if for no other reason than that the state public utility commission may request that it comply with it.

12.8 IDENTITY AND KEY MANAGEMENT

While authentication and cryptographic algorithms, which include public/private key schemes, have been discussed throughout this book, we believe it is important to emphasize the critical importance that key management will play in both reliably identifying devices and encrypting their information. It has become an article of faith within the cryptography community that encryption and digital signatures do not solve security problems; they simply allow organizations to shift the problem to a location where the problem can be better addressed. That is where key management comes in.

Because a digital key, whether it is in symmetric or asymmetric form, is what is ultimately used to unlock an encrypted message, uniquely sign a message to prove their authorship and authenticity, and verify the signature, the strength of cryptography is only as good as how these digital keys are protected.[5] Moreover, this is only effective if all the pieces can talk with each other. For example, a distribution management system may want to verify the identity of a message sent from a meter. Under many AMI (advanced metering infrastructure) implementations, key management stops at the head-end system, which often just dumps the meter data into an eXtensible Markup Language (XML) file that is retrieved by the meter data management system that is then read by the distribution management system. At this point, any digital signatures or encryption is stripped off. While they may be perfectly acceptable under some scenarios, some utilities may want that extra signature data available. Additionally, many would prefer to leverage their directory services, where public keys and other organizational data are frequently stored. This means that AMI, distribution automation, and other Smart Grid device vendors should avoid implementing proprietary key management systems where possible and develop their products so that centralized key management can be deployed. This has the advantage of providing better end-to-end security and does not inhibit the utility from future growth. Because certificate standards such as X.509v3 are well established for public key infrastructures (PKIs), implementing cryptographic algorithms on devices to support this format should be easy. Additionally, having that flexibility also makes it easier to support stronger cryptographic algorithms for encryption, hashing, and digital signatures as the algorithms evolve.

Ultimately, it is up to the utilities to define their architectures in a way that is open and remind vendors of the importance of supporting open architectures. This also means that utilities must think past their next forklift upgrade and take a more holistic and strategic approach to how all the pieces fit together and how they can interact with customers, service providers, and maybe even regulators. This can be quite a challenge as the inclination is to let product vendors define the solution for a given function and simply pick the best value for that function without looking at the bigger picture. Simply ask product vendors about how they implement key management and how it interoperates with the enterprise. If that question is asked enough, the solutions will be there.

12.9 DIFFERENTIAL POWER ANALYSIS AND OTHER SIDE CHANNEL ATTACKS

It may seem odd that such a seemingly obscure subject would find its way into a broad-based book about Smart Grid security, but this is not as obscure as one might think. Differential power analysis (DPA) involves taking very precise measurements of energy emanating from a device, usually over the air, in order to extract normally highly protected information. Using commercially available toolkits, hackers can extract cryptographic keys stored on a device from up to 200 feet away. For example, in workshops offered by Cryptography Research,[6] students with limited or no experience in the field are able to extract cryptographic keys from mobile phones after just a few hours of training using the toolkits provided. The reason why this is particularly relevant to Smart Grid technology is that a vast array of meters, collectors, line sensors, and other field devices are directly exposed to the public; and unlike mobile devices, they remain in one place with no one around to notice someone with a funny-looking device trying to measure power consumption.

However, the good news is that there are defenses. For decades, the military and intelligence communities have been building secure facilities and products using what are known as TEMPEST shielding, which is a broad category of technology and techniques to shield a device or facility in order to minimize electromagnetic emanations. Similarly, Cryptography Research offers several patented technologies to protect against DPA at the chip level that can be adopted by semiconductor manufacturers to significantly reduce the chances that an attacker will extract sensitive information using DPA.

However, DPA is one of a category of attacks known as side channel attacks that include electromagnetic emanations, vibrations, and the timing of computations. Many of these are difficult to perform outside a controlled lab environment, but they are nonetheless worthy of investigation. One that is particularly relevant to service providers and those providing cloud computing services is a co-tenancy or co-residency attack. This involves an attacker running his software on the same hardware as his target and collecting information about the processor, memory, hard drive, and cache in order to collect information on the target application or virtual machine, the term for a virtualized computer system that is typically one of many running on the same piece of hardware. In most cases, being able to place one's virtual machine on the same piece of hardware as the intended target is difficult enough in a large cloud computing environment. Actually, extracting useful information after co-tenancy is established is much more difficult.[7]

The point of these examples is not to scare anyone away from Smart Grid with advanced attacks that seem expensive to counter. Instead, they are designed to encourage creative thinking to solve challenges that are not insurmountable. More importantly, they provide utilities with a justification to force product vendors to really think about security and offer the types of robust protections that the industry needs.

12.10 ENERGY THEFT AND MARKET MANIPULATION

When asked why he robs banks, Willie Sutton, the famous bank robber, was reported to have said, "Because that's where the money is." And while he may never have said that phrase, the observation is no less valid. Notwithstanding threats from terrorists, nation-states, and bored teenagers, threats to the Smart Grid will come fast and furious when there is money to be made. In some senses, Smart Grid was intended to combat that problem, and to some extent it does. Italy's primary motivation for deploying smart metering was to cut down on energy theft. And where marijuana growers are jury-rigging electrical lines from the nearest transformer into their growing houses or where Third-World kids make a living climbing up on poles to give their neighbors electricity, smart metering can certainly help. However, cyber security threats may end up displacing the physical security threats that smart metering helped curtail. As noted previously, if someone is able to update the meter's firmware with a rogue version that reports electricity usage at half its actual level, then the problem will not be solved but will be exacerbated as the rogue copy of the firmware and the security exploit to apply it will spread over the Internet faster than a bootleg copy of the latest Lady Gaga track.

Of course, energy theft at the residential level is a relatively small-time endeavor. Cutting one's electricity bill is only going to net someone a few hundred dollars a month. And even then, the risk of detection is going to be high for significant reductions, so shaving off more than 20 to 30 percent may not be a wise decision. Of greater value to the professional criminal would be market manipulation. The introduction of the Smart Grid significantly increases the opportunities for the creation of financial markets at different stages in the electricity delivery system. For example, charging a battery during off-peak times when energy is cheaper and selling it back during peak times is classic "buy low, sell high" economic behavior. There will undoubtedly be energy exchanges where these transactions can take place, and like all markets there will be middlemen taking transaction fees as well as opportunities to buy options and derivatives. In those situations, hackers could potentially exfiltrate market-sensitive information that could give parties an advantage in a transaction. The growth of distributed generation will likely spawn a whole host of exchanges in local and national markets. Every piece of the Smart Grid architecture could potentially serve as a data point in support of financial transactions. That may also mean that logging and record management systems will need to be improved and perhaps subject to additional regulations such as the Sarbanes-Oxley Act, insider trading laws, and other financial industry rules.

12.11 PRIVACY

While one would hope that it will not be the case, it seems inevitable that privacy will become part of the discussions around Smart Grid. After all, the Netherlands suspended its Smart Grid deployment for more than a year as a result of privacy concerns and finally allowed it only after requiring utilities to give customers the option to opt out of having a smart meter.[8] While privacy is often a subjective issue, situations like this, in our opinion, represent an extreme response to a problem that is often more about public relations than about legitimate privacy concerns. Utilities should appreciate the fact that if customers are taken for granted and are not convinced that they stand to benefit from Smart Grid deployments, these kinds of responses will be common. More importantly, however, are the legitimate privacy considerations. Consumers should be given the option to grant or deny consent to sharing their energy data with third parties, particularly for marketing purposes. Government agencies should not be able to perform fishing expeditions on terabytes of energy use data to identify potential violations of the laws. A legitimate law enforcement purpose paired with a subpoena that is narrowly construed should continue to be the standard. Utilities are relatively new to data mining techniques, particularly in the area of marketing. The practice offers a tremendous opportunity for growth, but utilities should seek to learn from the financial services and retail industries about how to best exploit these opportunities without consumers feeling exploited.

12.12 WILL THE SMART GRID BE SECURE?

After reading this book, it seems fair to ask whether all these efforts can be done securely within the budgets allotted. Unfortunately, there is no hard and fast answer. Security is a journey, not a destination. Nothing will ever be 100 percent secure. Life is full of risks. It's just most of everyday risks like car accidents, violent crime, and severe weather are things that we've internalized and have come to accept with a few precautions thrown in. It is the risks that we do not understand well that tend to give us the most heartburn and often expect experts in such fields to provide easy answers. Nonetheless, utilities and vendors can focus on some key principles when rolling out the Smart Grid. They are

- *Use a strategic approach.* Utilities and the vendors that support them should understand the roadmap for Smart Grid even if all the details have not been ironed out. That means knowing all the systems that today's AMI system will be talking to in the future, over what protocols, and using what infrastructure. Utilities may need to make investment decisions based on available resources. However, they also need to ask what other new functions will be leveraging the same investments and whether they will be adequate in the future.
- *Cross organizational boundaries.* In many utilities, organizational elements are siloed, with little interaction between the different components. This is often the case with information technology folks supporting the enterprise and back-office functions who have little interaction with the power engineers supporting the distribution, transmission, and generation functions. Because Smart Grid is bringing those pieces together, it is critical that those parts have mutual respect for each other and share their knowledge in order for security to be implemented appropriately.
- *Become compliant through lower risk.* As stated numerous times in this book, simply complying with laws and regulations related to cyber security will not be enough to avoid being compromised. The threats are constantly changing, but security standards, laws, and regulations move slowly. Utilities must be aware of the current state of the threats and respond to them regardless of whether or not they are legally obligated to do so. In the end, staying on top of risks will also address the vast majority of compliance requirements.
- *Take a holistic approach to security.* Many people new to cyber security see advertisements for firewalls, anti-virus, or other security tools and assume that buying a particular technology will make them secure. That approach rarely works. Technology only works as well as the people using it. Many of today's security threats begin as a social engineering attack, such as a seemingly innocent e-mail message asking the user to click on a link or open an attachment that may even appear to come from someone they know. Without appropriate training and a healthy dose of skepticism, organizations are doomed to a vicious cycle of compromise and remediation. Ensuring that the organization combines equal doses of people, process, and technology when approaching security challenges makes it more likely that the risks will be mitigated to the greatest degree possible.

- *Be future-aware.* It is a common goal with technology to be future-proof. Unfortunately, that is not always possible. Products released today could not only be rendered obsolete by some newer product, but the older product could be rendered completely unusable based on changes to an infrastructure underlying communication protocols or data formats. With respect to Smart Grid, the goal should seek to be aware that changes will happen and to do one's best to ensure that the technologies and processes used today will work when changes occur in the future. For example, having the ability to remotely update smart meter firmware may be a critical feature given the likelihood of performance- or security-related patches that will need to be applied. Physically visiting each meter will certainly not be feasible.

And so the answer to the question about whether the Smart Grid can be secure is that it cannot be with any absolute certainty. However, if the steps above are followed, along with the other guidance we have provided throughout this book, the Smart Grid can reduce risk to a level that is acceptable for the vast majority of utilities, regulators, and consumers. That is the best that we can do.

ENDNOTES

1. Clarke, Richard and Knake, Robert, *Cyber War: The Next Threat to National Security and What to Do About It,* Ecco, New York, 2010.
2. Clarke, Richard and Knake, Robert, *Cyber War: The Next Threat to National Security and What to Do About It,* Ecco, New York, 2010, p. 5.
3. See McAfee Foundstone Professional Services and McAfee Labs, Global Energy Cyberattacks: Night Dragon, McAfee, February 10, 2011, p. 3, retrieved from http://www.mcafee.com/us/resources/white-papers/wp-global-energy-cyberattacks-night-dragon.pdf. ("Starting in November 2009, coordinated covert and targeted cyberattacks have been conducted against global oil, energy, and petrochemical companies. These attacks have involved social engineering, spearphishing attacks, exploitation of Microsoft Windows operating systems vulnerabilities, Microsoft Active Directory compromises, and the use of remote administration tools (RATs) in targeting and harvesting sensitive competitive proprietary operations and project-financing information with regard to oil and gas field bids and operations. We have identified the tools, techniques, and network activities used in these continuing attacks — which we have dubbed Night Dragon — as originating primarily in China.")

4. Electricity Grid Modernization: Progress Being Made on Cyber Security Guidelines, but Key Challenges Remain to Be Addressed, Government Accountability Office, Washington, D.C., January 2011, p. 19, retrieved from http://www.gao.gov/new.items/d11117.pdf.

5. For a better understanding of public, or asymmetric, key cryptography and how it differs from symmetric key cryptography, see Introduction to Public Key Technology and the Federal PKI Infrastructure, NIST Special Publication 800-32, February 26, 2011, retrieved from http://csrc.nist.gov/publications/nistpubs/800-32/sp800-32.pdf. Additionally, for recommendations on how to implement an effective key management system, see Recommendation for Key Management, NIST Special Publication 800-57, March 2007, retrieved from http://csrc.nist.gov/publications/nistpubs/800-57/sp800-57-Part1-revised2_Mar08-2007.pdf.

6. See, for example, a recent workshop offered at the company's San Francisco office at http://www.cryptography.com/newsevents/press_releases/2011/03/18/workshop-20110426-27.html. More about DPA and available protections can be found at http://www.cryptography.com/technology/dpa.html.

7. For a complete description of the issue and an example of how this attack was successfully launched on Amazon's cloud environment, see Thomas Ristenpart, Eran Tromer, Hovav Shacham, and Stefan Savage, Hey, you, get off of my cloud: Exploring information leakage in third-party compute clouds, *Proceedings of CCS 2009*, ACM Press, Nov. 2009, pp. 199–212.

8. See Štajnarová, Monica, Data privacy and security in smart meters: How to face this challenge, *Workshop on Regulatory Aspects of Data Transmission, Data Security and Data Protection in Relation to Smart Metering*, Florence, Italy, November 26, 2010, pp. 9–10, retrieve from http://www.florence-school.eu/portal/page/portal/FSR_HOME/ENERGY/Policy_Events/Workshops/2010/Smart_Metering/Presentation_Stanjarova.pdf.

Bibliography

2-Day Training, April 26–27, 2011: Power Analysis Attacks and Countermeasures for AT Defense Applications. Cryptography Research, 2011, retrieved from http://www.cryptography.com/news-events/press_releases/2011/03/18/workshop-20110426-27.html.

About APS – Power Plants. APS, retrieved from http://www.aps.com/general_info/aboutAPS_18.html.

Archer, Christina L. and Mark Z. Jacobson. Supplying baseload power and reducing transmission requirements by interconnecting wind farms. *Journal of Applied Meteorology and Climatology,* November 2007, pp. 1701–1717, retrieved from http://www.stanford.edu/group/efmh/winds/aj07_jamc.pdf.

ARRA of 2009 Related to Western. Western Area Power Administration, http://www.wapa.gov/recovery/default.htm.

Barker, Elaine, William Barker, William Burr, William Polk, and Miles Smid, Recommendation for Key Management, NIST Special Publication 800-57, March 2007, retrieved from http://csrc.nist.gov/publications/nistpubs/800-57/sp800-57-Part1-revised2_Mar08-2007.pdf.

Bartley, William H., Life Cycle Management of Utility Transformer Assets, The Hartford Steam Bender, Klaus, No Health Threat from Smart Meters, Utilities Telecom Council, retrieved from http://www.smartgridnews.com/artman/uploads/1/NoSmartMeterHealthThreat_1_.pdf.

Boiler Inspection and Insurance Company, 2002, retrieved from http://www.bplglobal.net/eng/knowledge-center/download.aspx?id=196.

Bonneville Power Administration website, retrieved from http://www.bpa.gov.

Burn Plus Wet Scrubber for Exhaust Gas Cleaning. Crystec Technology Trading GmbH, retrieved from http://www.crystec.com/ksiburne.htm.

Capacity and Energy Emergencies. Standard EOP-002-3, NERC, retrieved from http://www.nerc.com/files/EOP-002-3.pdf.

Chowdhury, S., S.P. Chowdhury, and P. Crossley, *Microgrids and Active Distribution Networks.* The Institution of Engineering and Technology (London), 2009.

Clarke, Richard and Robert Knake, *Cyber War: The Next Threat to National Security and What to Do About It*, Ecco (New York), 2010.

Critical Cyber Asset Identification, Standard CIP-002-3, Cyber Security, NERC, retrieved from http://www.nerc.com/files/CIP-002-3.pdf.

Daniels, A. and W. Salter, What is SCADA? In the Proceedings of 7th International Conference on Accelerator and Large Experimental Physics Control Systems (ICALEPCS 99), Trieste, Italy, October 4–8, 1999, pp. 339–343.

Dhameja, Sandeep, *Electric Vehicle Battery Systems*. Newnes (Boston) 2002.

Directory: Cents per Kilowatt-Hour. Pure Energy Systems, April 13, 2011, retrieved from http://peswiki.com/index.php/Directory:Cents_Per_Kilowatt-Hour.

DPA Countermeasures. Cryptography Research, 2011, retrieved from http://www.cryptography.com/technology/dpa.html.

Du, Yangbo and John E. Parsons, Update on the Cost of Nuclear Power. Center for Energy and Environmental Policy Research, May 2009, retrieved from http://web.mit.edu/ceepr/www/publications/workingpapers/2009-004.pdf.

Edge Node 7000 Series. Echelon, 2011, retrieved from http://www.echelon.com/metering/ecn.htm.

Electric Power Research Institute, *Electricity Sector Framework for the Future, Volume I: Achieving the 21st Century Transformation*, Electric Power Research Institute (Washington, DC), 2003.

Electricity Basic Statistics – 100 Largest Electric Plants. U.S. Energy Information Administration, June 2010, retrieved from http://www.eia.doe.gov/neic/rankings/plantsbycapacity.htm.

Electricity Grid Modernization: Progress Being Made on Cyber Security Guidelines, but Key Challenges Remain to Be Addressed, GAO-11-117, Government Accountability Office, January 2011, retrieved from http://www.gao.gov/new.items/d11117.pdf.

Energy Policy Act (102nd Congress H.R.776.ENR).

Evans, James W., The Interface between Automation and the Substation, in *Electric Power Substations Engineering*, Chapter 6, pp. 1–20, edited by John D. McDonald, CRC Press (Boca Raton, FL), 2003.

Fact Sheet on Nuclear Insurance and Disaster Relief Funds. U.S. Nuclear Regulatory Commission, retrieved from http://www.nrc.gov/reading-rm/doc-collections/fact-sheets/funds-fs.html.

Falliere, Nicholas, Liam O. Murchu, and Eric Chien, W32.Stuxnet Dossier, Symantec Security Response, Version 1.4, February 2011, retrieved from http://www.symantec.com/content/en/us/enterprise/media/security_response/whitepapers/w32_stuxnet_dossier.pdf.

Federal Energy Management Program Using Distributed Energy Resources, A How-To Guide for Federal Facility Managers, DOE/GO-102002-1520, U.S. Department of Energy by the National Renewable Energy Laboratory, May 2002.

Fehrenbacher, Katie, 5 Energy Storage Players That Won Smart Grid Stimulus Funds. Gigom, November 24, 2009, retrieved from http://gigaom.com/cleantech/5-energy-storage-players-that-won-smart-grid-stimulus-funds/.

Fehrenbacher, Katie, Lesson Learned from the PGandE Smart Meter Suit: It's a Communication Problem. GigaOM, November 19, 2009, retrieved from http://gigaom.com/cleantech/lesson-learned-from-the-pge-smart-meter-suit-its-a-communication-problem/.

Fehrenbacher, Katie, Smart Meter Backlash, Again: This Time in Texas. GigaOM, March 10, 2010, retrieved from http://gigaom.com/cleantech/smart-meter-backlash-again-this-time-in-texas/.

Flick, Tony and Justin Morehouse, *Securing the Smart Grid: Next Generation Power Grid Security,* Syngress (Burlington, MA), 2011.

Gabriel, Mark A., *Visions for a Sustainable Energy Future.* Fairmont Press (Lilburn, GA), 2008.

Glossary of Terms Used in Reliability Standards. NERC, April 20, 2009, retrieved from http://www.nerc.com/files/Glossary_2009April20.pdf.

Glossary of Terms Used in Reliability Standards. NERC, February 12, 2008, retrieved from http://www.nerc.com/files/Glossary_12Feb08.pdf.

Gorman, Siobhan, Electricity Grid in U.S. Penetrated by Spies. *Wall Street Journal,* April 8, 2009, retrieved from http://online.wsj.com/article/SB123914805204099085.html.

Herold, Rebecca, Privacy Challenges of the Smart Power Grid. Rebecca Herold and Associates, LLC, 2011.

HomePlug AV standard. RF Design, August 2006, pp. 16–26, retrieved from http://rfdesign.com/mag/608RFDF1.pdf.

Husain, Iqbal, *Electric and Hybrid Vehicles: Design Fundamentals, 2nd ed.,* CRC Press (Boca Raton, FL), 2011.

Ice Bear Energy Storage System. Ice Bear Energy Storage, retrieved from http://www.ice-energy.com/ice-bear-energy-storage-system.

IEC 60870-6-503, Part 6-503, Telecontrol Protocols Compatible with ISO Standards and ITU-T Recommendations: TASE.2 Services and Protocol, 2nd edition, 2002–2004.

Fries, Steffen, Hans Joachim Hof, and Maik Seewald, Enhancing IEC 62351 to Improve Security for Energy Automation in Smart Grid Environments, *2010 Fifth International Conference on Internet and Web Applications and Services,* IEEE, 2010, pp. 135–142.

IEC 61968, Part 1, Interface Architecture and General Requirements, International Electrotechnical Commission, 2003.

IEC 61970, Energy Management System Application Program Interface (EMS-API), Part 301: Common Information Model (CIM) Base, International Electrotechnical Commission, 2009.

IEC 62351, Parts 1–8, *Information Security for Power System Control Operations,* International Electrotechnical Commission, 2006–2008.

Kanellos, Michael, Can a Country Get 90 Percent of Its Power From Renewables? Wired, April 3, 2011, retrieved from http://www.wired.com/epicenter/2011/04/country-90pct-renewables/.

Katar, Srinivas, et. al. Harnessing the potential of powerline communications using the Homeland Security Presidential Directive / HSPD-7, The White House, December 17, 2003, retrieved from http://www.fas.org/irp/offdocs/nspd/hspd-7.html.

Kuhn, D. Richard, Vincent C. Hu, W. Timothy Polk, and Shu-Jen Chang, Introduction to Public Key Technology and the Federal PKI Infrastructure, NIST Special Publication 800-32, February 26, 2011, retrieved from http://csrc.nist.gov/publications/nistpubs/800-32/sp800-32.pdf.

Kuphaldt, R., Lessons in Electric Circuits. Volume IV (Digital), January 18, 2006, retrieved from http://www.ibiblio.org/kuphaldt/electric-Circuits/Digital/index.html.

LaMonica, Martin, PGandE admits to flaws in some smart meters, *CNet News,* May 11, 2010, retrieved from http://news.cnet.com/8301-11128_3-20004645-54.html.

LaMonica, Martin, Start-up GridGlo taps smart-meter data deluge. *CNet News,* May 11, 2011, retrieved from http://news.cnet.com/8301-11128_3-20061749-54.html?part=rssandsubj=newsand tag=2547-1_3-0-20.

LAN/MAN Standards Committee of the IEEE Computer Society, Part 15.4: Wireless Medium Access Control (MAC) and Physical Layer (PHY) Specifications for Low-Rate Wireless Personal Area Networks (WPANs), IEEE-SA Standards Board, June 2006, retrieved from http://standards.ieee.org/getieee802/download/802.15.4-2006.pdf.

Layton, Lyndsey, Reliance on Coal Sullies "Green the Capitol" Effort. *Washington Post,* April 27, retrieved from http://www.washington-post.com/wp-dyn/content/article/2007/04/20/AR2007042002128.html.

Letendre, Steven, Denholm, Paul, Lilienthal, Peter, Electric and Hybrid Cars, New Load or New Resource? *Public Utilities Fortnightly,* December 2006, pp. 28–37.

Majdalawieh, Munir, *DNPSec:* Distributed Network Protocol Version 3 (DNP3) Security Framework, December 19, 2005, retrieved from http://www.acsac.org/2005/techblitz/majdalawieh.pdf.

McAfee Foundstone Professional Services and McAfee Labs, Global Energy Cyberattacks: Night Dragon, McAfee, February 10, 2011, retrieved from http://www.mcafee.com/us/resources/white-papers/wp-global-energy-cyberattacks-night-dragon.pdf.

McNaughton, Gary A. and Robert Saint, Comparison of the MultiSpeak® Distribution Connectivity Model and the IEC Common Information Model Network Data Set, 2008, retrieved from http://www.multi-speak.org/documents/MultiSpeak_and_CIM_DIstributech_2008_article.pdf.

MODBUS Protocol, Modbus Organization, Inc., 2011, retrieved from http://www.modbus.org/specs.php.

Montenegro, G., N. Kushalnagar, J. Hui, and D. Culler, Transmission of IPv6 Packets over IEEE 802.15.4 Networks, September 2007, retrieved from http://tools.ietf.org/html/rfc4944.

Morioka, Matthew, Alireza Abrishamkar, and Yve Kay, Three Gorges Dam, retrieved from http://www.eng.hawaii.edu/~panos/444_09_4_9.pdf.

The Need for Essential Consumer Protections: Smart Metering Proposals and the Move to Time-Based Pricing, AARP, National Consumer Law Center, National Association of State Utility Consumer Advocates, Consumers Union, and Public Citizen, August 2010, retrieved from http://www.nclc.org/images/pdf/energy_utility_tele-com/additional_resources/adv_meter_protection_report.pdf.

The NETL Smart Grid Implementation Strategy (SGIS). National Energy Technology Laboratory, 2011, retrieved from http://www.netl.doe.gov/smartgrid/.

Newborough, M. and P. Augood, Demand-side management opportunities for the UK domestic sector, *IEE Proceedings of Generation Transmission and Distribution*, 146 (3) (1999), pp. 283–293.

NIST Framework and Roadmap for Smart Grid Interoperability Standards, Release 1.0, NIST Special Publication 1108, January 2010, retrieved from http://collaborate.nist.gov/twiki-sggrid/pub/SmartGrid/IKBFramework/NISTFrameworkAndRoadmapForSmartGridInteroperability_Release1final.pdf.

No Name Key – In the News. retrieved from http://www.nonamekey.org/nnknews.html.

North American Electric Reliability Corporation, retrieved from http://www.nerc.com/.

Nuqui, Reynaldo Francisco, State Estimation and Voltage Security Monitoring Using Synchronized Phasor Measurements, dissertation submitted to the Faculty of the Virginia Polytechnic Institute and State University, July 2, 2001 (Blacksburg, VA), retrieved from http://scholar.lib.vt.edu/theses/available/etd-07122001-030152/unrestricted/rnuqui_dissertation.pdf.

Paris, Demetrius and F. Kenneth Hurd, *Basic Electromagnetic Theory*, McGraw-Hill (New York), 1969.

Plug-In Hybrid Electric Vehicle RandD Plan, U.S. Department of Energy, retrieved from http://www1.eere.energy.gov/vehiclesandfules/pdfs/program/phev_rd_plan_02-28-07.pdf.

Rexford, Kenneth B., Peter R. Giuliani, and Leo Chartrand, *Electrical Control for Machines, Sixth edition,* Delmar Cengage Learning, 2002.

San Diego Gas and Electric Company Schedule Statin Power Self-Supply (SPSS), Feb. 4, 2010, retrieved from http://www.sdge.com/tm2/pdf/ELEC_ELEC-SCHEDS_SPSS.pdf.

Ristenpart, Thomas, Eran Tromer, Hovav Shacham, and Stefan Savage, Hey, you, get off of my cloud: Exploring information leakage in third-party compute clouds. *Proceedings of CCS 2009,* ACM Press, November 2009, pp. 199–212.

Schneier, Bruce, The Psychology of Security. January 18, 2008, retrieved from http://www.schneier.com/essay-155.html.

Security Profile for Distribution Management, Version 0.12, The Advanced Security Acceleration Project for the Smart Grid (ASAP-SG), August 16, 2010, retrieved from http://www.smartgridipedia.org/images/1/1b/DM_Security_Profile_-_v0_12_-_20100816.pdf.

Taum Sauk Pumped Storage Project (No. P-2277), Dam Breach Incident. FERC, June 28, 2010, retrieved from http://www.ferc.gov/industries/hydropower/safety/projects/taum-sauk.asp.

Technical Conference on Smart Grid Interoperability Standards (RM11-2-000) (Washington, DC), FERC, January 31, 2011 (submitted testimony), retrieved from http://ferc.gov/EventCalendar/EventDetails.aspx?ID=5571andCalType=%20andCalendarID=116andDate=01/31/2011andView=Listview.

Smart Grid Interoperability Panel – Cyber Security Working Group Standards Review, Phase 1 Report, October 7, 2010, found at http://collaborate.nist.gov/twiki-sggrid/pub/SmartGrid/CSCTGStandards/StandardsReviewPhase-1Report.pdf.

Southeastern Power Administration website, retrieved from http://www.sepa.doe.gov/.

Southwestern Power Administration website, retrieved from http://www.swpa.gov/.

Štajnarová, Monica, Data privacy and security in smart meters: How to face this challenge? *Workshop on Regulatory Aspects of Data Transmission, Data Security and Data Protection in Relation to Smart Metering,* Florence, November 26, 2010, pp. 9-10, retrieved from http://www.florence-school.eu/portal/page/portal/FSR_HOME/ENERGY/Policy_Events/Workshops/2010/Smart_Metering/Presentation_Stanjarova.pdf.

Standard BAL-002-1 – Disturbance Control Performance. NERC, August 5, 2010 retrieved from http://www.nerc.com/files/BAL-002-WECC-1.pdf.

Standard EOP-001-0 – Emergency Operations Planning. NERC, April 1, 2005, retrieved from http://www.nerc.com/files/EOP-001-0.pdf.

State Security Breach Notification Laws. National Conference of State Legislatures, October 12, 2010, retrieved from http://www.ncsl.org/default.aspx?tabid=13489.

TransFlow 2000: Lowest Cost Utility-scale Energy Storage. Premium Power, retrieved from http://www.premiumpower.com/product/transflow2000.php.

Transmission Vegetation Management Program, Standard FAC-003-1, NERC, retrieved from http://www.nerc.com/files/FAC-003-1.pdf.

UCAIug Home Area Network System Requirements Specification. OpenHAN Task Force formed by the SG Systems Working Group under the Open Smart Grid (OpenSG) Technical Committee of the UCA® International Users Group, version 2.0, August 30, 2010, retrieved from http://osgug.ucaiug.org/sgsystems/openhan/Shared%20Documents/OpenHAN%202.0/UCAIug%20HAN%20SRS%20-%20v2.0.pdf.

United States of America Federal Energy Regulatory Commission, 18 CFR Part 35, Docket No. RM99-2-000; Order No. 2000, Regional Transmission Organizations. December 20, 1999.

Energy Provisions in the American Recovery and Reinvestment Act of 2009 (P.L. 111-5), March 3, 2009.

Useful Thermal Output by Energy Source by Combined Heat and Power Producers. U.S. Energy Information Administration, April 2011, retrieved from http://www.eia.doe.gov/cneaf/electricity/epa/epat2p2.html.

Watertown Operations Center. Western Area Power Administration, retrieved from http://www.wapa.gov/newsroom/WTbrochure.htm.

Western Area Power Administration website, retrieved from http://www.wapa.gov.

Wang, Shimo and George Rodriguez, Smart RAS (Remedial Action Scheme), Southern California Edison (SCE), retrieved from http://asset.sce.com/Documents/Environment%20-%20Smart%20Grid/SmartRemedialActionScheme.pdf.

What FERC Does. FERC, retrieved from http://www.ferc.gov/about/ferc-does.asp.

What is Google PowerMeter? Google, 2011, retrieved from http://www.google.com/powermeter/about/about.html.

What is TED? Energy, Inc., 2010, retrieved from http://www.theenergydetective.com/about-ted.

Wright, Joshua, KillerBee: Practical ZigBee Exploitation Framework or "Wireless Hacking and the Kinetic World." retrieved from http://www.willhackforsushi.com/presentations/toorcon11-wright.pdf.

ZigBee® Membership, Designations and Logos Policy. ZigBee Alliance, January 27, 2009, retrieved from http://www.zigbee.org/imwp/idms/popups/pop_download.asp?ContentID=6700.

ZigBee Smart Energy Profile Specification, ZigBee Alliance, December 1, 2008, retrieved from http://zigbee.org/Standards/ZigBeeSmartEnergy/PublicApplicationProfile.aspx.

Index